Uncertainty and Operations Research

For further volumes:
http://www.springer.com/series/11709

Xiang Li

Credibilistic Programming

An Introduction to Models and Applications

 Springer

Xiang Li
Beijing University of Chemical Technology
Beijing, China, People's Republic

ISSN 2195-996X   ISSN 2195-9978 (electronic)
ISBN 978-3-642-36375-7   ISBN 978-3-642-36376-4 (eBook)
DOI 10.1007/978-3-642-36376-4
Springer Heidelberg New York Dordrecht London

Library of Congress Control Number: 2013935851

Printed on acid-free paper

Springer is part of Springer Science+Business Media (www.springer.com)

Preface

Fuzziness is a basic type of subjective uncertainty. The study on fuzziness was started in 1965 after the publication of Zadeh's seminal work *Fuzzy Sets*. In order to measure the chance of a fuzzy event occurs, Zadeh proposed the concepts of possibility measure and necessity measure, which are proved to be normal, nonnegative, monotone, but not self-dual. Since the duality is intuitive and important in both theory and practice, Liu and Liu (2002) defined a credibility measure as the average value of possibility measure and necessity measure, which was redefined by Li and Liu (2006) as a set function satisfying the normality, monotonicity, duality, and maximality axioms. After that, credibility measure was widely used in the fields of fuzzy decision, fuzzy process, fuzzy calculus, fuzzy differential equation, fuzzy logic, fuzzy inference, and so on. Nowadays, credibility theory has become a branch of mathematics for studying the behavior of fuzzy phenomena. Chapter 1 will be devoted to the credibility theory.

The decision analysis with fuzzy objective or fuzzy constraints is natural in some real-world applications, and sometimes seems to be inevitable. Credibilistic programming is a type of mathematical programming used to handle the fuzzy decision problems. In past years, researchers have proposed many efficient modeling approaches, which have been widely applied to many real-life problems. For example, Liu and Liu (2002) formulated an expected value model to minimize the average value of the objective under certain expected constraints. Liu and Iwamura (1998) proposed a maximax chance-constrained programming model, and Liu (1998) proposed a maximin chance-constrained programming model, which respectively maximizes the optimistic value and the pessimistic value of the objective with assumption that the fuzzy constraints will hold with certain confidence levels. Based on the concepts of fuzzy entropy, Li et al. (2011) proposed an entropy optimization model to minimize the uncertainty on possible values of the fuzzy objective, and Qin et al. (2009) formulated a cross-entropy minimization model to minimize the divergence of the fuzzy objective from a priori fuzzy quantity. Recently, Li et al. (2012) provided a regret minimization model to minimize the distance between the fuzzy objective values and the best values. Chapter 2 will provide a general introduction on the credibilistic programming as well as the genetic algorithm. Then the following chapters will respectively introduce the expected value model, chance-constrained programming model, entropy optimization model, cross-entropy minimization model, and regret minimization model.

The purpose of this book is to provide a powerful mathematical tool to handle the fuzzy decision problems. The book provides a self-contained and comprehensive presentation of credibilistic programming models and applications. The book is suitable for researchers, engineers, and students in the fields of management science, operations research, financial analysis, industrial engineering, information science, computer science, artificial intelligence, and so on. The readers will learn numerous new and efficient modeling ideas, and find this work a stimulating and useful reference.

Acknowledgment

This work was supported by the National Natural Science Foundation of China (No. 71101007), and the Specialized Research Fund for the Doctoral Program of Higher Education of China (No. 20110009120036).

Beijing University of Chemical Technology, China, People's Republic   Xiang Li
November 14, 2012

Contents

List of Frequently Used Symbols

Θ	Universal set
$\mathcal{A}$	Power set
Cr	Credibility measure
$(\Theta, \mathcal{A}, \text{Cr})$	Credibility space
ξ, η, τ	Fuzzy variables
$\boldsymbol{\xi}$	Fuzzy vector
ν, μ	Credibility functions
(a, b)	Equipossible fuzzy variable
(a, b, c)	Triangular fuzzy variable
(a, b, c, d)	Trapezoidal fuzzy variable
$N(e, \sigma)$	Normal fuzzy variable
$EXP(m)$	Exponential fuzzy variable
f	Objective function
g	Constraint function
$\boldsymbol{x}$	Decision vector
U	Credibilistic mapping
$E[\xi]$	Expected value of fuzzy variable ξ
$V[\xi]$	Variance of fuzzy variable ξ
$S[\xi]$	Skewness of fuzzy variable ξ
$\xi_{\sup}(\alpha)$	α-optimistic value of fuzzy variable ξ
$\xi_{\inf}(\alpha)$	α-pessimistic value of fuzzy variable ξ
$H[\xi]$	Entropy of fuzzy variable ξ
$D[\xi; \eta]$	Cross-entropy of ξ from η
$d(\xi, \eta)$	Distance between fuzzy variables ξ and η
$\emptyset$	Empty set
$\Re$	The set of real numbers
$\vee$	Maximum operator
$\wedge$	Minimum operator
$\forall$	Universal quantifier
$\exists$	Existential quantifier

Chapter 1
Credibility Theory

The concept of fuzzy set was initialized by Zadeh (1965) via membership function in 1965. In order to measure the chance of a fuzzy event occurs, Zadeh proposed the concepts of possibility measure (Zadeh 1978) and necessity measure (Zadeh 1979). It is proved that both possibility measure and necessity measure satisfy the properties of normality, nonnegativity and monotonicity. However, neither of them is self-dual. Since the duality is intuitive and important in both theory and practice, Liu and Liu (2002) defined a credibility measure as the average value of possibility measure and necessity measure, which was redefined by Li and Liu (2006a) as a set function satisfying the normality, monotonicity, duality, and maximality axioms. Nowadays, credibility measure has been well applied to many definitions for fuzzy variables, such as expected value (Liu and Liu 2002), variance (Liu 2004), skewness (Li et al. 2010a), independence (Li and Liu 2006b; Liu and Gao 2007), optimistic value and pessimistic value (Liu 2004), entropy (Li and Liu 2008a, 2007), cross-entropy (Qin et al. 2009), distance (Li and Liu 2008b), and so on. Credibility theory has become a branch of axiomatic mathematics for modeling fuzziness.

This chapter mainly introduces some basic concepts and important theorems including credibility measure, fuzzy variable, credibility function, independence, identical distribution, credibility subadditivity theorem, credibility semicontinuous theorem, credibility extension theorem, product credibility theorem, credibility inversion theorem, Zadeh extension theorem, and so on.

1.1 Credibility Measure

Let Θ be a nonempty set, and let $\mathcal{A}$ be its power set (i.e., the collection of all subsets). Each element of $\mathcal{A}$ is called an event. Credibility measure is a set function from $\mathcal{A}$ to $[0, 1]$. For each event, its credibility indicates the chance that the event will occur. In order to ensure that the set function has certain mathematical properties, Li and Liu (2006a) provided the following four axioms:

X. Li, *Credibilistic Programming*, Uncertainty and Operations Research,
DOI 10.1007/978-3-642-36376-4_1, © Springer-Verlag Berlin Heidelberg 2013

Axiom 1 (Normality) $\text{Cr}\{\Theta\} = 1$ *for the universal set* Θ.

Axiom 2 (Monotonicity) $\text{Cr}\{A\} \leq \text{Cr}\{B\}$ *for any events* $A \subseteq B$.

Axiom 3 (Duality) $\text{Cr}\{A\} + \text{Cr}\{A^c\} = 1$ *for any event* A.

Axiom 4 (Maximality) $\text{Cr}\{\cup_i A_i\} = \sup_i \text{Cr}\{A_i\}$ *for any collection of events* $\{A_i\}$ *with* $\sup_i \text{Cr}\{A_i\} < 0.5$.

Remark 1.1 The human thinking is always dominated by the duality. If someone says a proposition is true with possibility p, then all of us will think that the proposition is false with possibility $1 - p$. For example, if someone tells us that "Tom is tall with possibility 0.7", then we will think that "Tom is not tall with possibility 0.3".

Definition 1.1 A set function is called a credibility measure if it satisfies the normality, monotonicity, duality, and maximality axioms.

Remark 1.2 Credibility measure is a non-additive measure. For example, if $\text{Cr}\{A\} = 0.3$ and $\text{Cr}\{B\} = 0.4$, then it follows from the maximality axiom that $\text{Cr}\{A \cup B\} = 0.4$, which implies that

$$\text{Cr}\{A \cup B\} \neq \text{Cr}\{A\} + \text{Cr}\{B\}.$$

In fact, a credibility measure is additive if and only if Θ contains at most two points with nonzero credibilities. If there are three points $\theta_1, \theta_2, \theta_3$ with nonzero credibilities, we can prove that Cr is non-additive. Without loss of generality, we assume

$$\text{Cr}\{\theta_1\} \leq \text{Cr}\{\theta_2\} \leq 0.5 \leq \text{Cr}\{\theta_3\}.$$

It follows from the maximality axiom and the duality axiom that

$$\text{Cr}\{\theta_1\} + \text{Cr}\{\theta_2\} + \text{Cr}\{\theta_3\} > \text{Cr}\{\theta_1, \theta_2\} + \text{Cr}\{\theta_3\} = \text{Cr}\{\Theta\}$$

which implies that Cr is non-additive.

Example 1.1 Let $\Theta = \{\theta_1, \theta_2\}$, and let $\mathcal{A}$ be its power set including four events $\emptyset$, $\{\theta_1\}$, $\{\theta_2\}$ and Θ. Define a set function

$$\text{Cr}\{\emptyset\} = 0, \qquad \text{Cr}\{\theta_1\} = 0.4, \qquad \text{Cr}\{\theta_2\} = 0.6, \qquad \text{Cr}\{\Theta\} = 1.$$

Then Cr is a credibility measure because it satisfies the four axioms.

Example 1.2 Let $\Theta = \{\theta_1, \theta_2, \ldots\}$ be a countable set, and let $\mathcal{A}$ be its power set. Define a real function from Θ to $[0, 1]$ as follows,

$$\rho(\theta) = 1/2 - 1/(i + 1), \quad i = 1, 2, \ldots$$

We will prove that the set function

$$
\text{Cr}\{A\} =
\begin{cases}
\sup\limits_{\theta_i \in A} \rho(\theta_i), & \text{if } A \text{ is finite} \\
1 - \sup\limits_{\theta_i \in A^c} \rho(\theta_i), & \text{if } A \text{ is infinite}
\end{cases}
$$

is a credibility measure. Since the normality, monotonicity and maximality follow immediately from the definition, we only prove the duality. For any event A, the proof breaks down into three cases. If A is finite, then A^c is infinite and

$$
\text{Cr}\{A\} + \text{Cr}\{A^c\} = \sup_{\theta_i \in A} \rho(\theta_i) + 1 - \sup_{\theta_i \in A} \rho(\theta_i) = 1.
$$

Similarly, if A^c is finite, then A is infinite and

$$
\text{Cr}\{A\} + \text{Cr}\{A^c\} = 1 - \sup_{\theta_i \in A^c} \rho(\theta_i) + \sup_{\theta_i \in A^c} \rho(\theta_i) = 1.
$$

Otherwise, if both A and A^c are infinite, then we have

$$
\sup_{\theta_i \in A} \rho(\theta_i) = \sup_{\theta_i \in A^c} \rho(\theta_i) = 0.5
$$

which implies that $\text{Cr}\{A\} + \text{Cr}\{A^c\} = 1 - 0.5 + 1 - 0.5 = 1$.

Example 1.3 Let Θ be the unit open interval $(0, 1)$, and let $\mathcal{A}$ be its power set. Define a set function

$$
\text{Cr}\{A\} =
\begin{cases}
1, & \text{if } A = \Theta \\
0, & \text{if } A = \emptyset \\
0.5, & \text{otherwise.}
\end{cases}
$$

Then it is easy to prove that Cr satisfies the normality, monotonicity, duality and maximality axioms, that is, Cr is a credibility measure.

Theorem 1.1 *The empty set has a credibility measure zero, i.e.,* $\text{Cr}\{\emptyset\} = 0$.

Proof Since $\Theta = \emptyset \cup \Theta$, it follows from the normality axiom $\text{Cr}\{\Theta\} = 1$ and the duality axiom that

$$
\text{Cr}\{\emptyset\} = 1 - \text{Cr}\{\Theta\} = 0.
$$

The proof is complete.   □

Theorem 1.2 *For any event A, we have* $0 \le \text{Cr}\{A\} \le 1$.

Proof It follows immediately from the monotonicity axiom because $\text{Cr}\{\emptyset\} = 0$, $\text{Cr}\{\Theta\} = 1$ and $\emptyset \subseteq A \subseteq \Theta$. The proof is complete.   □

Theorem 1.3 *For any events A and B with* $\text{Cr}\{A \cup B\} \le 0.5$, *we have*

$$\text{Cr}\{A \cup B\} = \text{Cr}\{A\} \vee \text{Cr}\{B\}. \tag{1.1}$$

Proof We will prove the equality by using the reduction to absurdity. If it does not hold, it follows from the maximality axiom that

$$\text{Cr}\{A \cup B\} > \text{Cr}\{A\} \vee \text{Cr}\{B\} \ge 0.5.$$

The contradiction proves that the equality holds. The proof is complete.   □

Remark 1.3 Note that the condition $\text{Cr}\{A \cup B\} \le 0.5$ cannot be removed in Theorem 1.3. In fact, for any event A with $0 < \text{Cr}\{A\} < 1$, we have

$$\text{Cr}\{\Theta\} > \text{Cr}\{A\} \vee \text{Cr}\{A^c\}.$$

Remark 1.4 Assume $\alpha \le 0.5$. For any events A and B, if it is known that the union event takes a credibility α, then it follows from Theorem 1.3 that there is at least one event has credibility α, that is, $\text{Cr}\{A\} = \alpha$ or $\text{Cr}\{B\} = \alpha$. However, the converse may be not true.

Theorem 1.4 *For any events A and B with* $\text{Cr}\{A \cap B\} \ge 0.5$, *we have*

$$\text{Cr}\{A \cap B\} = \text{Cr}\{A\} \wedge \text{Cr}\{B\}. \tag{1.2}$$

Proof It follows from the duality axiom that $\text{Cr}\{A^c \cup B^c\} \le 0.5$ and

$$\begin{aligned}
\text{Cr}\{A \cap B\} &= 1 - \text{Cr}\{A^c \cup B^c\} = 1 - \text{Cr}\{A^c\} \vee \text{Cr}\{B^c\} \\
&= 1 - \left(1 - \text{Cr}\{A\}\right) \vee \left(1 - \text{Cr}\{B\}\right) \\
&= \text{Cr}\{A\} \wedge \text{Cr}\{B\}.
\end{aligned}$$

The proof is complete.   □

Remark 1.5 Note that the condition $\text{Cr}\{A \cap B\} \ge 0.5$ cannot be removed in Theorem 1.4. In fact, for any event A with $0 < \text{Cr}\{A\} < 1$, we have

$$\text{Cr}\{\emptyset\} < \text{Cr}\{A\} \wedge \text{Cr}\{A^c\}.$$

Theorem 1.5 (Credibility Subadditivity Theorem, Liu (2004)) *The credibility measure is subadditive. That is, for any events A and B, we have*

$$\text{Cr}\{A \cup B\} \le \text{Cr}\{A\} + \text{Cr}\{B\}. \tag{1.3}$$

In fact, the credibility measure is also countably subadditive.

Proof For any events A and B, if $\mathrm{Cr}\{A\} \vee \mathrm{Cr}\{B\} < 0.5$, then it follows from the maximality axiom that

$$\mathrm{Cr}\{A \cup B\} = \mathrm{Cr}\{A\} \vee \mathrm{Cr}\{B\} \leq \mathrm{Cr}\{A\} + \mathrm{Cr}\{B\}.$$

Otherwise, we have $\mathrm{Cr}\{A\} \vee \mathrm{Cr}\{B\} \geq 0.5$. Without loss of generality, we assume $\mathrm{Cr}\{A\} \geq 0.5$. It follows from the duality axiom that $\mathrm{Cr}\{A^c\} \leq 0.5$, which implies that

$$\begin{aligned}
\mathrm{Cr}\{A^c\} &= \mathrm{Cr}\{A^c \cap B\} \vee \mathrm{Cr}\{A^c \cap B^c\} \\
&\leq \mathrm{Cr}\{A^c \cap B\} + \mathrm{Cr}\{A^c \cap B^c\} \\
&\leq \mathrm{Cr}\{B\} + \mathrm{Cr}\{A^c \cap B^c\}.
\end{aligned}$$

Applying this inequality, we obtain

$$\begin{aligned}
\mathrm{Cr}\{A\} + \mathrm{Cr}\{B\} &= 1 - \mathrm{Cr}\{A^c\} + \mathrm{Cr}\{B\} \\
&\geq 1 - \mathrm{Cr}\{B\} - \mathrm{Cr}\{A^c \cap B^c\} + \mathrm{Cr}\{B\} \\
&= 1 - \mathrm{Cr}\{A^c \cap B^c\} \\
&= \mathrm{Cr}\{A \cup B\}.
\end{aligned}$$

The proof is complete.   □

Remark 1.6 According to the subadditivity theorem, it is easy to prove that credibility measure is null-additive. That is, if $\mathrm{Cr}\{A\} = 0$ or $\mathrm{Cr}\{B\} = 0$, then we have $\mathrm{Cr}\{A \cup B\} = \mathrm{Cr}\{A\} + \mathrm{Cr}\{B\}$. In other words, the credibility of an event remains unchanged if it is enlarged or reduced by another event with measure zero.

Theorem 1.6 (Liu 2004) *Let $\{A_i\}$ be a sequence of events with*

$$\lim_{i \to \infty} \mathrm{Cr}\{A_i\} = 0.$$

Then for any event B, we have

$$\lim_{i \to \infty} \mathrm{Cr}\{B \cup A_i\} = \lim_{i \to \infty} \mathrm{Cr}\{B \setminus A_i\} = \mathrm{Cr}\{B\}. \tag{1.4}$$

Proof For any events A_i and B, it follows from the monotonicity axiom and the subadditivity theorem that

$$\mathrm{Cr}\{B\} \leq \mathrm{Cr}\{B \cup A_i\} \leq \mathrm{Cr}\{B\} + \mathrm{Cr}\{A_i\},$$

and

$$\mathrm{Cr}\{B \setminus A_i\} \leq \mathrm{Cr}\{B\} \leq \mathrm{Cr}\{B \setminus A_i\} + \mathrm{Cr}\{A_i\}.$$

Letting $i \to \infty$. According to the squeeze theorem, we have

$$\lim_{i \to \infty} \text{Cr}\{B \cup A_i\} = \lim_{i \to \infty} \text{Cr}\{B \backslash A_i\} = \text{Cr}\{B\}.$$

The proof is complete.   □

Theorem 1.7 (Credibility Semicontinuous Theorem, Liu (2004)) *For any event sequence* $\{A_i\}$ *with limit* A, *we have*

$$\lim_{i \to \infty} \text{Cr}\{A_i\} = \text{Cr}\{A\} \tag{1.5}$$

if one of the following conditions is satisfied:

(a) $\lim_{i \to \infty} \text{Cr}\{A_i\} < 0.5$ *and* $A_i \uparrow A$;
(b) $\text{Cr}\{A\} \le 0.5$ *and* $A_i \uparrow A$;
(c) $\lim_{i \to \infty} \text{Cr}\{A_i\} > 0.5$ *and* $A_i \downarrow A$;
(d) $\text{Cr}\{A\} \ge 0.5$ *and* $A_i \downarrow A$.

Proof (a) Since $\{A_i\}$ is an increasing sequence, we have

$$\sup_i \text{Cr}\{A_i\} = \lim_{i \to \infty} \text{Cr}\{A_i\} < 0.5.$$

Then it follows from the maximality axiom that

$$\text{Cr}\{A\} = \text{Cr}\left\{\bigcup_i A_i\right\} = \sup_i \text{Cr}\{A_i\} = \lim_{i \to \infty} \text{Cr}\{A_i\}.$$

(b) If $\lim_{i \to \infty} \text{Cr}\{A_i\} < 0.5$, it follows immediately from conclusion (a). Otherwise, according to the monotonicity axiom, we have

$$\text{Cr}\{A\} = 0.5 = \lim_{i \to \infty} \text{Cr}\{A_i\}.$$

(c) According to the duality axiom, we have

$$\lim_{i \to \infty} \text{Cr}\left\{A_i^c\right\} < 0.5.$$

Then it follows from conclusion (a) that

$$\text{Cr}\left\{A^c\right\} = \lim_{i \to \infty} \text{Cr}\left\{A_i^c\right\}.$$

Again, it follows from the duality axiom that

$$\text{Cr}\{A\} = \lim_{i \to \infty} \text{Cr}\{A_i\}.$$

(d) Based on the duality axiom, we have $\mathrm{Cr}\{A^c\} < 0.5$. Then it follows from conclusion (b) that

$$\mathrm{Cr}\{A\} = 1 - \mathrm{Cr}\{A^c\} = 1 - \lim_{i \to \infty} \mathrm{Cr}\{A_i^c\} = \lim_{i \to \infty} \mathrm{Cr}\{A_i\}.$$

The proof is complete. □

Example 1.4 Generally speaking, the credibility measure is neither lower semicontinuous nor upper semicontinuous. Let us reconsider Example 1.3. Define a decreasing sequence of events $A_i = (0, 1/i]$ with

$$\lim_{i \to \infty} A_i = \bigcap_{i=1}^{\infty} (0, 1/i] = \emptyset.$$

It is easy to prove that

$$\lim_{i \to \infty} \mathrm{Cr}\{A_i\} = 0.5 > 0 = \mathrm{Cr}\{\emptyset\}.$$

On the other hand, define an increasing sequence of events $B_i = (0, 1 - 1/i]$ for $i = 1, 2, \ldots$. Since $\Theta = (0, 1)$, we have

$$\lim_{i \to \infty} B_i = \bigcup_{i=1}^{\infty} (0, 1 - 1/i] = \Theta.$$

However, it is easy to prove that

$$\lim_{i \to \infty} \mathrm{Cr}\{B_i\} = 0.5 < 1 = \mathrm{Cr}\{\Theta\}.$$

Theorem 1.8 (Credibility Asymptotic Theorem, Liu (2004)) *For any event sequence* $\{A_i\}, i = 1, 2, \ldots,$ *we have*

$$\lim_{i \to \infty} \mathrm{Cr}\{A_i\} \geq 0.5, \quad \textit{if } A_i \uparrow \Theta, \tag{1.6}$$

$$\lim_{i \to \infty} \mathrm{Cr}\{A_i\} \leq 0.5, \quad \textit{if } A_i \downarrow \emptyset. \tag{1.7}$$

Proof First, assume that $\{A_i\}$ is an increasing sequence with limit Θ. If

$$\lim_{i \to \infty} \mathrm{Cr}\{A_i\} < 0.5,$$

it follows from the credibility semicontinuous theorem that

$$\mathrm{Cr}\{\Theta\} = \lim_{i \to \infty} \mathrm{Cr}\{A_i\} < 0.5,$$

which is in contradiction with the normality axiom. Thus the first inequality is proved. On the other hand, if $A_i \downarrow \emptyset$, we have $A_i^c \uparrow \Theta$. Then it follows from (1.6) and the duality axiom that

$$\lim_{i \to \infty} \text{Cr}\{A_i\} = 1 - \lim_{i \to \infty} \text{Cr}\{A_i^c\} \leq 0.5.$$

The proof is complete.   □

For a nonempty set Θ, suppose that the credibility value of each singleton set is given. Is the credibilitymeasure fully and uniquely determined? The credibility extension theorem will answer this question. First, we introduce the credibility extension condition.

Theorem 1.9 (Credibility Extension Condition, Li and Liu (2006a)) *Suppose that Θ is a nonempty set. If* Cr *is a credibility measure, then we have*

$$\sup_{\theta \in \Theta} \text{Cr}\{\theta\} \geq 0.5, \tag{1.8}$$

$$\text{Cr}\{\theta^*\} + \sup_{\theta \neq \theta^*} \text{Cr}\{\theta\} = 1 \quad \textit{if } \text{Cr}\{\theta^*\} \geq 0.5. \tag{1.9}$$

Proof We first prove condition (1.8) by using the reduction to absurdity. If it does not hold, it follows from the maximality axiom that $\text{Cr}\{\Theta\} < 0.5$, which contradicts to the normality axiom. Therefore, we have

$$\sup_{\theta \in \Theta} \text{Cr}\{\theta\} \geq 0.5.$$

Suppose that θ^* is a point with $\text{Cr}\{\theta^*\} \geq 0.5$. It follows from the duality axiom and the maximality axiom that

$$\sup_{\theta \neq \theta^*} \text{Cr}\{\theta\} = \text{Cr}\{\Theta \backslash \{\theta^*\}\} = 1 - \text{Cr}\{\theta^*\}.$$

The proof is complete.   □

Remark 1.7 For a credibility measure, it follows from the duality axiom that there is at most one point with credibility value larger than 0.5. Therefore, the credibility extension condition has the following equivalent form

$$\sup_{\theta_1 \neq \theta_2} \left(\text{Cr}\{\theta_1\} + \text{Cr}\{\theta_2\} \right) = 1. \tag{1.10}$$

Especially, if the universal set Θ contains only two points θ_1 and θ_2, then the credibility extension condition essentially tells us that $\text{Cr}\{\theta_1\} + \text{Cr}\{\theta_2\} = 1$.

Remark 1.8 Let Θ be a nonempty set containing three points θ_1, θ_2 and θ_3, and let $\mathcal{A}$ be its power set. Define a set function on $\mathcal{A}$ satisfying

$$M\{\theta_1\} = 0.6, \qquad M\{\theta_2\} = 0.3, \qquad M\{\theta_3\} = 0.2.$$

Then M is not a credibility measure since it does not satisfy the credibility extension condition.

Theorem 1.10 (Credibility Extension Theorem, Li and Liu (2006a)) *Suppose that Θ is a nonempty set, and $\mathcal{A}$ is its power set. If $\mathrm{Cr}\{\theta\}$ is a nonnegative function satisfying the credibility extension condition, then it has a unique extension to a credibility measure on $\mathcal{A}$ as follows,*

$$\mathrm{Cr}\{A\} = \begin{cases} \sup_{\theta \in A} \mathrm{Cr}\{\theta\}, & \text{if } \sup_{\theta \in A} \mathrm{Cr}\{\theta\} < 0.5 \\ 1 - \sup_{\theta \in A^c} \mathrm{Cr}\{\theta\}, & \text{if } \sup_{\theta \in A} \mathrm{Cr}\{\theta\} \geq 0.5. \end{cases} \tag{1.11}$$

Proof We first prove that the set function (1.11) is a credibility measure, i.e., it satisfies the normality, monotonicity, duality and maximality axioms.

Step 1: Since $\mathrm{Cr}\{\theta\}$ satisfies the credibility extension condition, we have

$$\sup_{\theta \in \Theta} \mathrm{Cr}\{\theta\} \geq 0.5.$$

Then it follows from (1.11) that

$$\mathrm{Cr}\{\Theta\} = 1 - \sup_{\theta \in \emptyset} \mathrm{Cr}\{\theta\} = 1 - 0 = 1$$

which implies that the set function Cr satisfies the normality axiom.

Step 2: This step will prove that Cr satisfies the monotonicity axiom. For any events $A \subseteq B$, we have $B^c \subseteq A^c$, which implies that

$$\sup_{\theta \in A} \mathrm{Cr}\{\theta\} \leq \sup_{\theta \in B} \mathrm{Cr}\{\theta\}, \qquad \sup_{\theta \in B^c} \mathrm{Cr}\{\theta\} \leq \sup_{\theta \in A^c} \mathrm{Cr}\{\theta\}.$$

The following argument breaks down into two cases. If we have

$$\sup_{\theta \in A} \mathrm{Cr}\{\theta\} < 0.5,$$

then it follows from (1.11) that

$$\mathrm{Cr}\{A\} = \sup_{\theta \in A} \mathrm{Cr}\{\theta\} \leq \sup_{\theta \in B} \mathrm{Cr}\{\theta\} \leq \mathrm{Cr}\{B\}.$$

Otherwise, we have

$$\sup_{\theta \in B} \mathrm{Cr}\{\theta\} \geq \sup_{\theta \in A} \mathrm{Cr}\{\theta\} \geq 0.5.$$

For this case, it follows from (1.11) that

$$\mathrm{Cr}\{A\} = 1 - \sup_{\theta \in A^c} \mathrm{Cr}\{\theta\} \le 1 - \sup_{\theta \in B^c} \mathrm{Cr}\{\theta\} = \mathrm{Cr}\{B\}.$$

Step 3: This step will prove that Cr satisfies the duality axiom. For any event A, according to the credibility extension condition, we have

$$\sup_{\theta \in A} \mathrm{Cr}\{\theta\} \vee \sup_{\theta \in A^c} \mathrm{Cr}\{\theta\} \ge 0.5, \qquad \sup_{\theta \in A} \mathrm{Cr}\{\theta\} + \sup_{\theta \in A^c} \mathrm{Cr}\{\theta\} \le 1.$$

In order to prove $\mathrm{Cr}\{A\} + \mathrm{Cr}\{A^c\} = 1$, the argument breaks down into three cases. First, if we have

$$\sup_{\theta \in A} \mathrm{Cr}\{\theta\} < 0.5, \qquad \sup_{\theta \in A^c} \mathrm{Cr}\{\theta\} \ge 0.5,$$

it follows from (1.11) that

$$\mathrm{Cr}\{A\} + \mathrm{Cr}\{A^c\} = \sup_{\theta \in A} \mathrm{Cr}\{\theta\} + 1 - \sup_{\theta \in A} \mathrm{Cr}\{\theta\} = 1.$$

Similarly, if we have

$$\sup_{\theta \in A^c} \mathrm{Cr}\{\theta\} < 0.5, \qquad \sup_{\theta \in A} \mathrm{Cr}\{\theta\} \ge 0.5,$$

it follows from (1.11) that

$$\mathrm{Cr}\{A\} + \mathrm{Cr}\{A^c\} = 1 - \sup_{\theta \in A^c} \mathrm{Cr}\{\theta\} + \sup_{\theta \in A^c} \mathrm{Cr}\{\theta\} = 1.$$

Otherwise, we have

$$\sup_{\theta \in A} \mathrm{Cr}\{\theta\} = \sup_{\theta \in A^c} \mathrm{Cr}\{\theta\} = 0.5.$$

In this case, it follows from (1.11) that $\mathrm{Cr}\{A\} + \mathrm{Cr}\{A^c\} = 0.5 + 0.5 = 1$.

Step 4: This step will prove that Cr satisfies the maximality axiom. For any event collection $\{A_i\}$ with $\sup_i \mathrm{Cr}\{A_i\} < 0.5$, we have

$$\mathrm{Cr}\left\{\bigcup_i A_i\right\} = \sup_{\theta \in \bigcup_i A_i} \mathrm{Cr}\{\theta\} = \sup_i \sup_{\theta \in A_i} \mathrm{Cr}\{\theta\} = \sup_i \mathrm{Cr}\{A_i\}.$$

Thus Cr is a credibility measure because it satisfies the four axioms.

Now, let us prove the uniqueness. Assume that there is another credibility measure M which satisfies $M\{\theta\} = \mathrm{Cr}\{\theta\}$ for each $\theta \in \Theta$. We will prove that $M\{A\} = \mathrm{Cr}\{A\}$ for each event A. First, if we have

$$\sup_{\theta \in A} \mathrm{Cr}\{\theta\} < 0.5,$$

it follows from the maximality axiom and (1.11) that

$$M\{A\} = \sup_{\theta \in A} \mathrm{Cr}\{\theta\} = \mathrm{Cr}\{A\}.$$

Similarly, if we have

$$\sup_{\theta \in A^c} \mathrm{Cr}\{\theta\} < 0.5,$$

we can prove $M\{A^c\} = \mathrm{Cr}\{A^c\}$. Then it follows from the duality axiom that

$$M\{A\} = 1 - M\{A^c\} = 1 - \mathrm{Cr}\{A^c\} = \mathrm{Cr}\{A\}.$$

Finally, based on the credibility extension condition, we have

$$\sup_{\theta \in A} \mathrm{Cr}\{\theta\} = \sup_{\theta \in A^c} \mathrm{Cr}\{\theta\} = 0.5,$$

which implies that $\mathrm{Cr}\{A\} = 0.5$. On the other hand, according to the monotonicity axiom, we have $M\{A\} \geq 0.5$ and $M\{A^c\} \geq 0.5$. Furthermore, it follows from the duality axiom that $M\{A\} = 0.5$. The uniqueness is proved. □

Remark 1.9 Based on the credibility extension theorem, we can define a credibility measure by giving the credibility values of all singleton sets.

Example 1.5 Let Θ be the unit interval $[0, 1]$, and let $\mathcal{A}$ be its power set. Based on the credibility extension theorem, we can define a credibility measure by constructing a nonnegative real function satisfying the credibility extension condition. For example, we define $\mathrm{Cr}\{\theta\} = \theta/2$ for all $\theta \in [0, 1]$. Then it can be uniquely extended to a credibility measure

$$\mathrm{Cr}\{A\} = \begin{cases} \displaystyle\sup_{\theta \in A}(\theta/2), & \text{if } \sup_{\theta \in A} \theta < 1 \\ 1 - \displaystyle\sup_{\theta \in A^c}(\theta/2), & \text{if } \sup_{\theta \in A} \theta = 1 \end{cases}$$

for any $A \in \mathcal{A}$.

1.2 Credibility Space

This section introduces the concept of credibility space and the product credibility theorem.

Definition 1.2 Let Θ be a nonempty set, $\mathcal{A}$ the power set, and Cr a credibility measure. Then the triplet $(\Theta, \mathcal{A}, \mathrm{Cr})$ is called a credibility space.

Example 1.6 The triplet $(\Theta, \mathcal{A}, \mathrm{Cr})$ is a credibility space if $\Theta = \{\theta_1, \theta_2, \ldots\}$, $\mathcal{A}$ is the power set of Θ, and

$$\mathrm{Cr}\{\theta_i\} = 0.5, \quad i = 1, 2, \ldots \tag{1.12}$$

Note that for each $A \in \mathcal{A}$, the credibility measure is produced by using the credibility extension theorem as follows,

$$\mathrm{Cr}\{A\} = \begin{cases} 0, & \text{if } A = \emptyset \\ 1, & \text{if } A = \Theta \\ 0.5, & \text{otherwise.} \end{cases}$$

Definition 1.3 Suppose that $(\Theta_k, \mathcal{A}_k, \mathrm{Cr}_k)$, $k = 1, 2, \ldots, n$ are credibility spaces. Let $\Theta = \Theta_1 \times \Theta_2 \times \cdots \times \Theta_n$, and let $\mathcal{A}$ be the power set of Θ. The product credibility measure Cr is the credibility measure satisfying

$$\mathrm{Cr}\{\theta\} = \mathrm{Cr}_1\{\theta_1\} \wedge \mathrm{Cr}_2\{\theta_2\} \wedge \cdots \wedge \mathrm{Cr}_n\{\theta_n\} \tag{1.13}$$

for each $\theta = (\theta_1, \theta_2, \ldots, \theta_n)$.

Theorem 1.11 (Product Credibility Theorem) *The function* $\mathrm{Cr}\{\theta\}$ *defined by* (1.13) *satisfies the credibility extension condition.*

Proof For each $k = 1, 2, \ldots, n$, since $\mathrm{Cr}_k\{\theta_k\}$ satisfies the credibility extension condition, we have

$$\begin{aligned} \sup_{\theta \in \Theta} \mathrm{Cr}\{\theta\} &= \sup_{(\theta_1, \theta_2, \ldots, \theta_n) \in \Theta} \min_{1 \leq k \leq n} \mathrm{Cr}_k\{\theta_k\} \\ &= \min_{1 \leq k \leq n} \sup_{\theta_k \in \Theta_k} \mathrm{Cr}_k\{\theta_k\} \\ &\geq 0.5. \end{aligned}$$

Suppose that $\theta^* = (\theta_1^*, \theta_2^*, \ldots, \theta_n^*)$ is a point satisfying $\mathrm{Cr}\{\theta^*\} \geq 0.5$. Without loss of generality, let i be the index such that

$$\mathrm{Cr}\{\theta^*\} = \min_{1 \leq k \leq n} \mathrm{Cr}_k\{\theta_k^*\} = \mathrm{Cr}_i\{\theta_i^*\}.$$

For each k, according to the credibility extension condition on Cr_k, we have

$$\mathrm{Cr}_k\{\theta_k^*\} \geq 0.5, \tag{1.14}$$

$$\sup_{\theta_k \neq \theta_k^*} \mathrm{Cr}_k\{\theta_k\} \leq 0.5, \tag{1.15}$$

$$\sup_{\theta_i \neq \theta_i^*} \mathrm{Cr}_i\{\theta_i\} \geq \sup_{\theta_k \neq \theta_k^*} \mathrm{Cr}_k\{\theta_k\}. \tag{1.16}$$

Then it follows from inequalities (1.14)–(1.16) that

$$\sup_{\theta \neq \theta^*} \text{Cr}\{\theta\} = \sup_{\theta \neq \theta^*} \min_{1 \leq k \leq n} \text{Cr}_k\{\theta_k\}$$

$$= \sup_{1 \leq j \leq n} \sup_{\theta_j \neq \theta_j^*} \left(\text{Cr}_j\{\theta_j\} \wedge \min_{k \neq j} \text{Cr}_k\{\theta_k^*\} \right)$$

$$= \sup_{\theta_i \neq \theta_i^*} \left(\text{Cr}_i\{\theta_i\} \wedge \min_{k \neq i} \text{Cr}_k\{\theta_k^*\} \right)$$

$$= \sup_{\theta_i \neq \theta_i^*} \text{Cr}_i\{\theta_i\}.$$

Therefore, by using the credibility extension condition on Cr_i, we have

$$\text{Cr}\{\theta^*\} + \sup_{\theta \neq \theta^*} \text{Cr}\{\theta\} = \text{Cr}_i\{\theta_i^*\} + \sup_{\theta_i \neq \theta_i^*} \text{Cr}_i\{\theta_i\} = 1.$$

The proof is complete. □

Definition 1.4 Assume that $(\Theta_k, \mathcal{A}_k, \text{Cr}_k)$, $k = 1, 2, \ldots, n$ are credibility spaces. Let $\Theta = \Theta_1 \times \Theta_2 \times \cdots \times \Theta_n$, $\mathcal{A}$ the power set of Θ, and Cr the product credibility measure. Then the triplet $(\Theta, \mathcal{A}, \text{Cr})$ is called the product credibility space.

Remark 1.10 In what follows, we assume that all fuzzy variables are defined on the same credibility space. Otherwise, we may embed them into the product credibility space.

1.3 Fuzzy Variable

Roughly speaking, a fuzzy variable is a real function defined on a credibility space. See Fig. 1.1. A formal definition is given as follows.

Definition 1.5 (Liu 2004) A fuzzy variable is a function ξ from a credibility space $(\Theta, \mathcal{A}, \text{Cr})$ to the set of real numbers.

Example 1.7 Take a credibility space $(\Theta, \mathcal{A}, \text{Cr})$ to be $\{\theta_1, \theta_2\}$ with $\text{Cr}\{\theta_1\} = \text{Cr}\{\theta_2\} = 0.5$. Then the function

$$\xi(\theta) = \begin{cases} 0, & \text{if } \theta = \theta_1 \\ 1, & \text{if } \theta = \theta_2 \end{cases}$$

is a fuzzy variable.

Fig. 1.1 A fuzzy variable

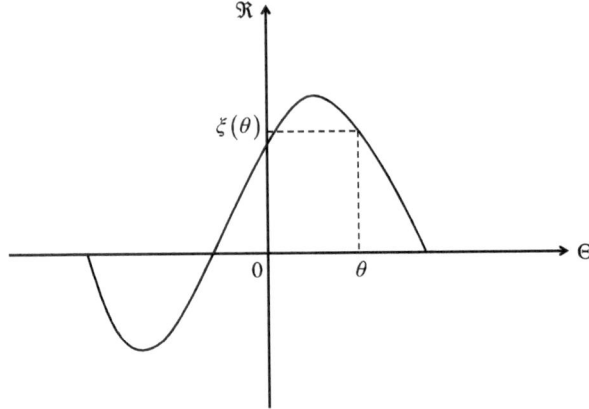

Example 1.8 Take a credibility space $(\Theta, \mathcal{A}, \mathrm{Cr})$ to be the closed interval $[0, 1]$ with $\mathrm{Cr}\{\theta\} = \theta/2$ for all $\theta \in [0, 1]$. Then the identify function

$$\xi(\theta) = \theta, \quad \forall \theta \in [0, 1]$$

is a fuzzy variable.

Example 1.9 A real number c may be regarded as a special fuzzy variable. In fact, it is the constant function $\xi(\theta) \equiv c$ on any credibility space $(\Theta, \mathcal{A}, \mathrm{Cr})$.

Remark 1.11 Suppose that ξ is a fuzzy variable defined on credibility space $(\Theta, \mathcal{A}, \mathrm{Cr})$. For any Borel set B of real numbers, it is easy to prove that

$$\{\xi \in B\} = \big\{\theta \in \Theta \mid \xi(\theta) \in B\big\}$$

is an element of $\mathcal{A}$. Therefore, the fuzzy variable ξ is essentially a measurable function from $(\Theta, \mathcal{A}, \mathrm{Cr})$ to the set of real numbers.

Definition 1.6 (Liu 2004) A fuzzy variable ξ is said to be

(a) nonnegative if $\mathrm{Cr}\{\xi < 0\} = 0$;
(b) positive if $\mathrm{Cr}\{\xi \le 0\} = 0$;
(c) nonpositive if $\mathrm{Cr}\{\xi > 0\} = 0$;
(d) negative if $\mathrm{Cr}\{\xi \ge 0\} = 0$;
(e) continuous if $\mathrm{Cr}\{\xi = x\}$ is a continuous function of x;
(f) simple if there is a finite sequence $\{x_1, x_2, \ldots, x_n\}$ such that

$$\mathrm{Cr}\{\xi \ne x_1, \xi \ne x_2, \ldots, \xi \ne x_n\} = 0;$$

(g) discrete if there is a countable sequence $\{x_1, x_2, \ldots\}$ such that

$$\mathrm{Cr}\{\xi \ne x_1, \xi \ne x_2, \ldots\} = 0.$$

Definition 1.7 Fuzzy variables ξ and η are said to be equal if and only if $\xi(\theta) = \eta(\theta)$ for all θ with nonzero credibility.

Definition 1.8 An n-dimensional fuzzy vector is a function from a credibility space $(\Theta, \mathcal{A}, \text{Cr})$ to the set of n-dimensional real vectors.

Theorem 1.12 *The vector* $\boldsymbol{\xi} = (\xi_1, \xi_2, \ldots, \xi_n)$ *is a fuzzy vector if and only if* $\xi_1, \xi_2, \ldots, \xi_n$ *are fuzzy variables.*

Proof If $\boldsymbol{\xi}$ is a fuzzy vector defined on the credibility space $(\Theta, \mathcal{A}, \text{Cr})$, then $\xi_1, \xi_2, \ldots, \xi_n$ are all functions from Θ to the set of real numbers. Thus they are fuzzy variables. Conversely, suppose that ξ_k, $k = 1, 2, \ldots, n$ are fuzzy variables on the credibility space $(\Theta, \mathcal{A}, \text{Cr})$. Then $\boldsymbol{\xi}$ is a function from $(\Theta, \mathcal{A}, \text{Cr})$ to the set of n-dimensional real vectors. Hence, $\boldsymbol{\xi}$ is a fuzzy vector. The theorem is proved. $\qquad\square$

Definition 1.9 Suppose that $f : \mathfrak{R}^n \to \mathfrak{R}$ is an n-dimensional function, and $\xi_1, \xi_2, \ldots, \xi_n$ are fuzzy variables on the credibility space $(\Theta, \mathcal{A}, \text{Cr})$. Then $\xi = f(\xi_1, \xi_2, \ldots, \xi_n)$ is a fuzzy variable defined as

$$\xi(\theta) = f\big(\xi_1(\theta), \xi_2(\theta), \ldots, \xi_n(\theta)\big), \quad \forall \theta \in \Theta. \tag{1.17}$$

Example 1.10 Let ξ_1 and ξ_2 be two fuzzy variables. Then (a) the sum $\xi = \xi_1 + \xi_2$ is a fuzzy variable defined by

$$\xi(\theta) = \xi_1(\theta) + \xi_2(\theta), \quad \forall \theta \in \Theta;$$

(b) the product $\xi = \xi_1 \cdot \xi_2$ is a fuzzy variable defined by

$$\xi(\theta) = \xi_1(\theta) \cdot \xi_2(\theta), \quad \forall \theta \in \Theta;$$

(c) the maximum $\xi = \xi_1 \vee \xi_2$ is a fuzzy variable defined by

$$\xi(\theta) = \xi_1(\theta) \vee \xi_2(\theta), \quad \forall \theta \in \Theta;$$

(d) the minimum $\xi = \xi_1 \wedge \xi_2$ is a fuzzy variable defined by

$$\xi(\theta) = \xi_1(\theta) \wedge \xi_2(\theta), \quad \forall \theta \in \Theta.$$

The reader may wonder whether ξ defined by (1.17) is a fuzzy variable. The following theorem answers this question.

Theorem 1.13 *Let* $\boldsymbol{\xi}$ *be an n-dimensional fuzzy vector, and let* $f : \mathfrak{R}^n \to \mathfrak{R}$ *be an n-dimensional function. Then* $f(\boldsymbol{\xi})$ *is a fuzzy variable.*

Proof Since $f(\boldsymbol{\xi})$ is a function from a credibility space to the set of real numbers, it is a fuzzy variable. The proof is complete. $\qquad\square$

Fig. 1.2 The shape of a
credibility function

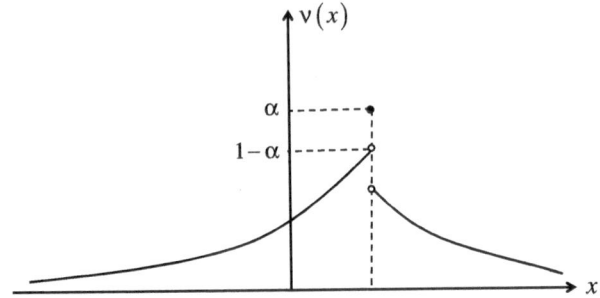

1.4 Credibility Function

For a fuzzy variable ξ, its credibility function is a mapping from $\Re$ to the set of unit interval $[0, 1]$ (see Fig. 1.2). For each $x \in \Re$, the credibility value represents the degree that the fuzzy variable takes value x.

Definition 1.10 Suppose that ξ is a fuzzy variable defined on the credibility space $(\Theta, \mathcal{A}, \mathrm{Cr})$. Then its credibility function is derived from the credibility measure as

$$v(x) = \mathrm{Cr}\{\xi = x\}, \quad \forall x \in \Re. \tag{1.18}$$

Theorem 1.14 *A real function* $v : \Re \to [0, 1]$ *is a credibility function for a fuzzy variable if and only if it satisfies*

$$\sup_{x \in \Re} v(x) \geq 0.5, \tag{1.19}$$

$$v(x^*) + \sup_{x \neq x^*} v(x) = 1, \quad \text{if } v(x^*) \geq 0.5.$$

Proof The necessity may be similarly proved with the credibility extension condition. Now, we prove the sufficiency. Assume that v is a real function satisfying condition (1.19). According to the credibility extension theorem, define a credibility space $(\Theta, \mathcal{A}, \mathrm{Cr})$ with $\Theta = \Re$ and $\mathrm{Cr}\{\theta\} = v(\theta)$ for all $\theta \in \Theta$. Define a fuzzy variable ξ as the identify function $\xi(\theta) = \theta$. For each $x \in \Re$, we have

$$\mathrm{Cr}\{\xi = x\} = v(x).$$

Therefore, v is a credibility function. The proof is complete. □

Theorem 1.15 *A continuous function* $v : \Re \to [0, 1]$ *is a credibility function if and only if it satisfies*

$$\sup_{x \in \Re} v(x) = 0.5. \tag{1.20}$$

Proof For a continuous function (see Fig. 1.3), it is easy to prove that condition (1.19) and condition (1.20) are equivalent. The proof is complete. □

Fig. 1.3 The shape of a
continuous credibility
function

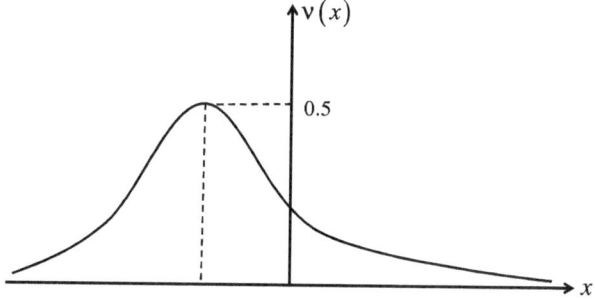

Example 1.11 Suppose that v is a simple function defined by

$$v(x) = \begin{cases} 0.2, & \text{if } x = x_1 \\ 0.3, & \text{if } x = x_2 \\ 0.7, & \text{if } x = x_3 \\ 0.3, & \text{if } x = x_4 \\ 0.2, & \text{if } x = x_5. \end{cases}$$

It follows from Theorem 1.14 that v is a credibility function.

Example 1.12 Suppose that v is a discrete function taking values in a countable set $\{x_1, x_2, \ldots\}$ with

$$v(x_i) = (i - 1)/2(i + 1), \quad i = 1, 2, \ldots$$

It follows from Theorem 1.14 that v is a credibility function.

Example 1.13 Let $v : [0, \pi] \to [0, 1]$ be a real function defined by $v(x) = \alpha \sin(x)$. If $\alpha = 0.5$, it follows from Theorem 1.15 that v is a credibility function since

$$\sup_{x \in \Re} v(x) = 0.5.$$

However, if $\alpha \neq 0.5$, then v is not a credibility function.

Theorem 1.16 (Credibility Inversion Theorem) *Let ξ be a fuzzy variable with credibility function v. Then for any set B of real numbers, we have*

$$\text{Cr}\{\xi \in B\} = \begin{cases} \sup_{x \in B} v(x), & \text{if } \sup_{x \in B} v(x) < 0.5 \\ 1 - \sup_{x \in B^c} v(x), & \text{if } \sup_{x \in B} v(x) \geq 0.5. \end{cases} \tag{1.21}$$

Proof It follows immediately from the credibility extension theorem. □

Fig. 1.4 Equipossible
credibility function

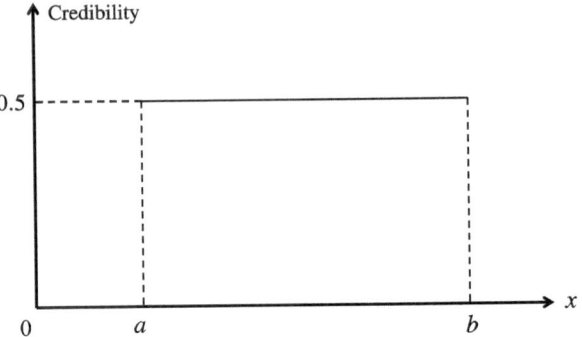

Example 1.14 Let ξ be a fuzzy variable defined by the credibility function

$$v(x) = \begin{cases} 0.3, & \text{if } x = 1 \\ 0.4, & \text{if } x = 2 \\ 0.6, & \text{if } x = 3 \\ 0.1, & \text{if } x = 4. \end{cases}$$

Define sets $A = [1, 2]$ and $B = [1, 3]$. Since $\sup_{x \in A} v(x) < 0.5$, it follows from the credibility inversion theorem that

$$\mathrm{Cr}\{\xi \in A\} = \sup_{x \in A} v(x) = 0.4.$$

On the other hand, since $\sup_{x \in B} v(x) = 0.6$, we have

$$\mathrm{Cr}\{\xi \in B\} = 1 - \sup_{x \in B^c} v(x) = 1 - 0.1 = 0.9.$$

Example 1.15 An *equipossible fuzzy variable* $\xi = (a, b)$ is defined by the following credibility function (see Fig. 1.4)

$$v(x) = \begin{cases} 0.5, & \text{if } a \leq x \leq b \\ 0, & \text{otherwise,} \end{cases} \tag{1.22}$$

which takes value from $[a, b]$ with the same possibility.

Example 1.16 A *triangular fuzzy variable* $\xi = (a, b, c)$ is defined by the following credibility function (see Fig. 1.5)

$$v(x) = \begin{cases} (x - a)/2(b - a), & \text{if } a \leq x \leq b \\ (c - x)/2(c - b), & \text{if } b < x \leq c \\ 0, & \text{otherwise.} \end{cases} \tag{1.23}$$

Fig. 1.5 Triangular
credibility function

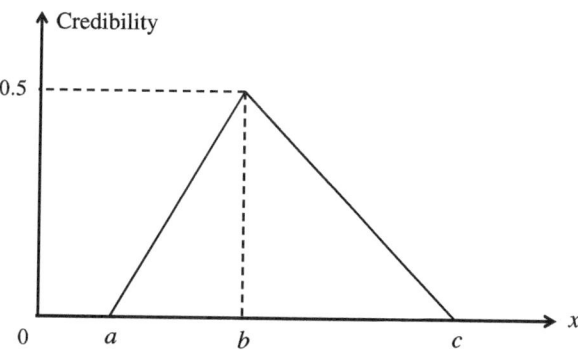

Fig. 1.6 Trapezoidal
credibility function

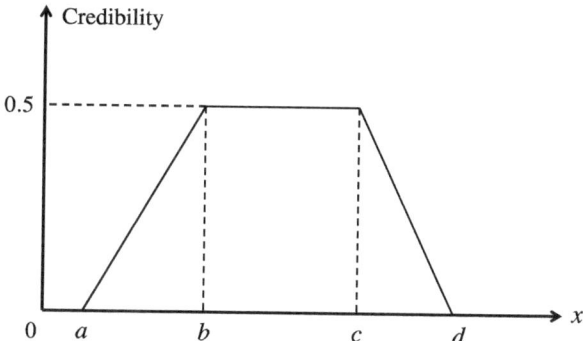

It is clear that a triangular fuzzy variable has a unimodal credibility function, which takes value b with the maximum credibility 0.5. Furthermore, a triangular fuzzy variable ξ will be called symmetric if $b - a = c - b$.

Example 1.17 A *trapezoidal fuzzy variable* $\xi = (a, b, c, d)$ is defined by the following credibility function (see Fig. 1.6)

$$\nu(x) = \begin{cases} (x - a)/2(b - a), & \text{if } a \leq x \leq b \\ 0.5, & \text{if } b < x \leq c \\ (d - x)/2(d - c), & \text{if } c < x \leq d \\ 0, & \text{otherwise.} \end{cases} \quad (1.24)$$

A trapezoidal fuzzy variable ξ will be called symmetric if $b - a = d - c$.

Example 1.18 For any $m > 0$, an *exponential fuzzy variable* $\xi = EXP(m)$ is defined by the following credibility function (see Fig. 1.7)

$$\nu(x) = 1/\big(1 + \exp\big(\pi x/(\sqrt{6}m)\big)\big), \quad x \geq 0.$$

It is clear that the exponential credibility function is decreasing and continuous on $[0, +\infty)$, which takes the maximum value 0.5 at zero and tends to zero as $x \to \infty$.

Fig. 1.7 Exponential
credibility function with
$m = 1$

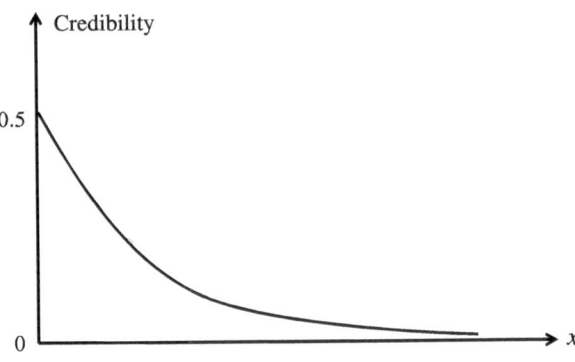

Fig. 1.8 Normal credibility
function with $e = 0$ and $\sigma = 1$

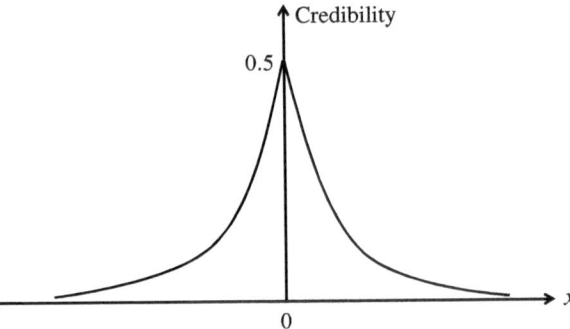

Example 1.19 For any $e \in \Re$ and $\sigma > 0$, a *normal fuzzy variable* $\xi = N(e, \sigma)$ is defined by the following credibility function (see Fig. 1.8)

$$\mu(x) = 1/\big(1 + \exp(\pi|x - e|/(\sqrt{6}\sigma))\big), \quad x \in \Re.$$

It is easy to prove that the normal credibility function is continuous, unimodal and symmetric with respect to parameter e.

Definition 1.11 Suppose that $(\xi_1, \xi_2, \ldots, \xi_n)$ is a fuzzy vector defined on the credibility space $(\Theta, \mathcal{A}, \mathrm{Cr})$. Then its joint credibility function is derived from the credibility measure as

$$\nu(x_1, x_2, \ldots, x_n) = \mathrm{Cr}\{\xi_1 = x_1, \xi_2 = x_2, \ldots, \xi_n = x_n\} \qquad (1.25)$$

for all $(x_1, x_2, \ldots, x_n) \in \Re^n$.

1.5 Independence

The fuzzy independence has been discussed by many authors from different angles, for example, De Cooman (1997), Hisdal (1978), Klir (1999), Li and Liu (2006b),

Liu and Gao (2007), Yager (1992), Zadeh (1975), and so on. In this section, we use the following definition.

Definition 1.12 The fuzzy variables $\xi_1, \xi_2, \ldots, \xi_m$ are said to be independent if and only if

$$\nu(x_1, x_2, \ldots, x_m) = \min_{1 \le i \le m} \nu(x_i) \tag{1.26}$$

for all $(x_1, x_2, \ldots, x_m) \in \Re^m$.

Theorem 1.17 *Fuzzy variables $\xi_1, \xi_2, \ldots, \xi_m$ are independent if and only if*

$$\mathrm{Cr}\{\xi_1 \in B_1, \xi_2 \in B_2, \ldots, \xi_m \in B_m\} = \min_{1 \le i \le m} \mathrm{Cr}\{\xi_i \in B_i\} \tag{1.27}$$

for any sets $B_1, B_2, \ldots, B_m$ of real numbers.

Proof Assume that fuzzy variables $\xi_1, \xi_2, \ldots, \xi_m$ are independent. For any sets $B_1, B_2, \ldots, B_m$ of real numbers, the proof of equation (1.27) breaks down into two cases.

Case 1. $\mathrm{Cr}\{\xi_1 \in B_1, \xi_2 \in B_2, \ldots, \xi_m \in B_m\} < 0.5$. In this case, according to the monotonicity axiom, we have

$$\sup_{x_i \in B_i, 1 \le i \le m} \nu(x_1, x_2, \ldots, x_m) < 0.5.$$

Furthermore, it follows from the maximality axiom that

$$\mathrm{Cr}\{\xi_1 \in B_1, \xi_2 \in B_2, \ldots, \xi_m \in B_m\}$$

$$= \sup_{x_i \in B_i, 1 \le i \le m} \nu(x_1, x_2, \ldots, x_m)$$

$$= \sup_{x_i \in B_i, 1 \le i \le m} \min_{1 \le i \le m} \nu_i(x_i)$$

$$= \min_{1 \le i \le m} \sup_{x_i \in B_i} \nu_i(x_i)$$

$$= \min_{1 \le i \le m} \mathrm{Cr}\{\xi_i \in B_i\}.$$

Case 2. $\mathrm{Cr}\{\xi_1 \in B_1, \xi_2 \in B_2, \ldots, \xi_m \in B_m\} \ge 0.5$. For this case, according to the duality axiom, we have

$$\mathrm{Cr}\left\{\left(\xi_1 \in B_1^c\right) \cup \left(\xi_2 \in B_2^c\right) \cup \cdots \cup \left(\xi_m \in B_m^c\right)\right\} \le 0.5.$$

Furthermore, it follows from Theorem 1.3 that

$$\mathrm{Cr}\{\xi_1 \in B_1, \xi_2 \in B_2, \ldots, \xi_m \in B_m\}$$

$$= 1 - \mathrm{Cr}\left\{\left(\xi_1 \in B_1^c\right) \cup \left(\xi_2 \in B_2^c\right) \cup \cdots \cup \left(\xi_m \in B_m^c\right)\right\}$$

$$= 1 - \max_{1 \le i \le m} \mathrm{Cr}\{\xi_i \in B_i^c\}$$

$$= \min_{1 \le i \le m} \mathrm{Cr}\{\xi_i \in B_i\}.$$

Conversely, assume that (1.27) holds for any sets $B_1, B_2, \dots, B_m$ of real numbers. Then for any $(x_1, x_2, \dots, x_m) \in \Re^m$, we have

$$\nu(x_1, x_2, \dots, x_m) = \mathrm{Cr}\{\xi_1 = x_1, \xi_2 = x_2, \dots, \xi_m = x_m\}$$

$$= \min_{1 \le i \le m} \mathrm{Cr}\{\xi_i = x_i\}$$

$$= \min_{1 \le i \le m} \nu_i(x_i),$$

which implies that fuzzy variables $\xi_1, \xi_2, \dots, \xi_m$ are independent. The proof is complete. $\qquad\square$

Theorem 1.18 *Fuzzy variables $\xi_1, \xi_2, \dots, \xi_m$ are independent if and only if*

$$\mathrm{Cr}\{(\xi_1 \in B_1) \cup (\xi_2 \in B_2) \cup \dots \cup (\xi_m \in B_m)\} = \max_{1 \le i \le m} \mathrm{Cr}\{\xi_i \in B_i\} \qquad (1.28)$$

for any sets $B_1, B_2, \dots, B_m$ of real numbers.

Proof If fuzzy variables $\xi_1, \xi_2, \dots, \xi_m$ are independent, it follows from Theorem 1.17 and the duality axiom that

$$\mathrm{Cr}\{(\xi_1 \in B_1) \cup (\xi_2 \in B_2) \cup \dots \cup (\xi_m \in B_m)\}$$

$$= 1 - \mathrm{Cr}\{\xi_1 \in B_1^c, \xi_2 \in B_2^c, \dots, \xi_m \in B_m^c\}$$

$$= 1 - \min_{1 \le i \le m} \mathrm{Cr}\{\xi_i \in B_i^c\}$$

$$= \max_{1 \le i \le m} \mathrm{Cr}\{\xi_i \in B_i\}.$$

Conversely, assume that (1.28) holds. For any sets $B_1, B_2, \dots, B_m$ of real numbers, it follows from the duality axiom that

$$\mathrm{Cr}\{\xi_1 \in B_1, \xi_2 \in B_2, \dots, \xi_m \in B_m\}$$

$$= 1 - \mathrm{Cr}\{(\xi_1 \in B_1^c) \cup (\xi_2 \in B_2^c) \cup \dots \cup (\xi_m \in B_m^c)\}$$

$$= 1 - \max_{1 \le i \le m} \mathrm{Cr}\{\xi_i \in B_i^c\}$$

$$= \min_{1 \le i \le m} \mathrm{Cr}\{\xi_i \in B_i\}.$$

According to Theorem 1.17, fuzzy variables $\xi_1, \xi_2, \dots, \xi_m$ are independent. The proof is complete. $\qquad\square$

Theorem 1.19 *Suppose that $\xi_1, \xi_2, \ldots, \xi_m$ are independent fuzzy variables. Then for any real functions $f_1, f_2, \ldots, f_m$, fuzzy variables $f_1(\xi_1), f_2(\xi_2), \ldots, f_m(\xi_m)$ are also independent.*

Proof For any $(x_1, x_2, \ldots, x_m) \in \Re^m$, according to Theorem 1.17, we have

$$\text{Cr}\{f_1(\xi_1) = x_1, f_2(\xi_2) = x_2, \ldots, f_m(\xi_m) = x_m\}$$

$$= \text{Cr}\{\xi_1 \in f_1^{-1}(x_1), \xi_2 \in f_2^{-1}(x_2), \ldots, \xi_m \in f_m^{-1}(x_m)\}$$

$$= \min_{1 \le i \le m} \text{Cr}\{\xi_i \in f_i^{-1}(x_i)\}$$

$$= \min_{1 \le i \le m} \text{Cr}\{f_i(\xi_i) = x_i\}.$$

Thus $f_1(\xi_1), f_2(\xi_2), \ldots, f_m(\xi_m)$ are independent. The proof is complete. $\square$

Example 1.20 If fuzzy variables ξ and η are independent, then it follows from above theorem that fuzzy variables ξ^2 and $(\eta + 1)^3$ are also independent.

Theorem 1.20 (Zadeh Extension Theorem) *Suppose that fuzzy variables $\xi_1, \xi_2, \ldots, \xi_m$ are independent. If f is a function from $\Re^m$ to $\Re$, then fuzzy variable $f(\xi_1, \xi_2, \ldots, \xi_m)$ has the credibility function*

$$\mu(z) = \begin{cases} \sup_{f(\boldsymbol{x})=z} \min_{1 \le i \le m} \nu_i(x_i), & \text{if } \sup_{f(\boldsymbol{x})=z} \min_{1 \le i \le m} \nu_i(x_i) < 0.5 \\ 1 - \sup_{f(\boldsymbol{x})\ne z} \min_{1 \le i \le m} \nu_i(x_i), & \text{if } \sup_{f(\boldsymbol{x})=z} \min_{1 \le i \le m} \nu_i(x_i) \ge 0.5 \end{cases} \quad (1.29)$$

for all $z \in \Re$, where ν_i is the credibility function of ξ_i for $i = 1, 2, \ldots, m$.

Proof Since fuzzy variables $\xi_1, \xi_2, \ldots, \xi_m$ are independent, the fuzzy vector $(\xi_1, \xi_2, \ldots, \xi_m)$ has a joint credibility function

$$\nu(\boldsymbol{x}) = \min_{1 \le i \le m} \nu_i(x_i), \quad \forall \boldsymbol{x} \in \Re^m.$$

For each $z \in \Re$, it follows from the credibility extension theorem that

$$\mu(z) = \begin{cases} \sup_{f(\boldsymbol{x})=z} \nu(\boldsymbol{x}), & \text{if } \sup_{f(\boldsymbol{x})=z} \nu(\boldsymbol{x}) < 0.5 \\ 1 - \sup_{f(\boldsymbol{x})\ne z} \nu(\boldsymbol{x}), & \text{if } \sup_{f(\boldsymbol{x})=z} \nu(\boldsymbol{x}) \ge 0.5. \end{cases}$$

The proof is complete. $\square$

Remark 1.12 The independence condition cannot be removed in Zadeh extension theorem. For example, define $\xi = (0, 1, 2)$ and $\eta = -\xi$. It is clear that $\xi + \eta = 0$. However, if we apply the Zadeh extension theorem, we have $\xi + \eta = (-2, 0, 2)$.

Example 1.21 Suppose that ξ is a nonnegative fuzzy variable with credibility function $\nu(x)$. Then for any $\alpha, \beta \in \Re$ with $\alpha \neq 0$, fuzzy variable $\alpha\xi + \beta$ has a credibility function

$$\mu(x) = \nu\big((x - \beta)/\alpha\big),$$

fuzzy variable $\sqrt[n]{\xi}$ has a credibility function

$$\mu(x) = \nu\big(x^n\big),$$

and fuzzy variable ξ^n has a credibility function

$$\mu(x) = \nu\big(\sqrt[n]{x}\big).$$

For example, if ξ is an equipossible fuzzy variable taking values in $[a, b]$, then we have $2\xi + 1 = (2a + 1, 2b + 1)$, $\xi^2 = (a^2, b^2)$, and $\sqrt{\xi} = (\sqrt{a}, \sqrt{b})$.

Example 1.22 Suppose that fuzzy variables $\xi_1 = (a_1, b_1)$ and $\xi_2 = (a_2, b_2)$ are independent with credibility functions ν_1 and ν_2, respectively. Then for any nonnegative real numbers α_1 and α_2, we have

$$\alpha_1\xi_1 + \alpha_2\xi_2 = (\alpha_1 a_1 + \alpha_2 a_2, \alpha_1 b_1 + \alpha_2 b_2). \tag{1.30}$$

For simplicity, we denote $a = \alpha_1 a_1 + \alpha_2 a_2$ and $b = \alpha_1 b_1 + \alpha_2 b_2$. We will prove that fuzzy variable $\alpha_1\xi_1 + \alpha_2\xi_2$ has the following credibility function

$$\nu(x) = \begin{cases} 0.5, & \text{if } a \leq x \leq b \\ 0, & \text{otherwise.} \end{cases}$$

When $\alpha_2 = 0$, the conclusion is trivial. Therefore, we assume that $\alpha_2 \neq 0$ in the following proof. The argument breaks down into three cases.

Case 1. $x < a$. In this case, for any $\alpha_1 x_1 + \alpha_2 x_2 = x$, we have $x_1 < a_1$ or $x_2 < a_2$. Then it follows from the Zadeh extension theorem that

$$\nu(x) = \sup_{\alpha_1 x_1 + \alpha_2 x_2 = x} \big(\nu_1(x_1) \wedge \nu_2(x_2)\big) = 0.$$

Case 2. $a \leq x \leq b$. Define $x_1^* = a_1$ and $x_2^* = (x - \alpha_1 a_1)/\alpha_2$. It is easy to prove that $a_2 \leq x_2^* \leq b_2$. It follows from the Zadeh extension theorem that

$$\nu(x) = \nu_1\big(x_1^*\big) \wedge \nu_2\big(x_2^*\big) = 0.5.$$

Case 3. $x > b$. For any $\alpha_1 x_1 + \alpha_2 x_2 = x$, we have $x_1 > b_1$ or $x_2 > b_2$. It follows from the Zadeh extension theorem that $\nu(x) = 0$.

Example 1.23 Suppose that triangular fuzzy variables $\xi_1 = (a_1, b_1, c_1)$ and $\xi_2 = (a_2, b_2, c_2)$ are independent with credibility functions ν_1 and ν_2, respectively. Then for any nonnegative real numbers α_1 and α_2, we have

$$\alpha_1\xi_1 + \alpha_2\xi_2 = (\alpha_1 a_1 + \alpha_2 a_2, \alpha_1 b_1 + \alpha_2 b_2, \alpha_1 c_1 + \alpha_2 c_2). \tag{1.31}$$

For simplicity, we denote $a = \alpha_1 a_1 + \alpha_2 a_2$, $b = \alpha_1 b_1 + \alpha_2 b_2$, and $c = \alpha_1 c_1 + \alpha_2 c_2$. If we use ν to denote the credibility function of fuzzy variable $\alpha_1 \xi_1 + \alpha_2 \xi_2$, we will prove that

$$\nu(x) = \begin{cases} 0, & \text{if } x < a \\ (x - a)/2(b - a), & \text{if } a \leq x < b \\ (c - x)/2(c - b), & \text{if } b \leq x < c \\ 0, & \text{if } x \geq c. \end{cases}$$

The proof is trivial if $\alpha_1 = 0$ or $\alpha_2 = 0$. In what follows, we assume that $\alpha_1 \alpha_2 \neq 0$. The argument breaks down into four cases.

Case 1. $x < a$. For any $\alpha_1 x_1 + \alpha_2 x_2 = x$, we have $x_1 < a_1$ or $x_2 < a_2$. Then it follows from the Zadeh extension principle that $\nu(x) = 0$.

Case 2. $a \leq x < b$. It follows from the Zadeh extension theorem that

$$\nu(x) = \sup_{a_1 \leq y_1 \leq b_1, a_2 \leq y_2 \leq b_2, \alpha_1 y_1 + \alpha_2 y_2 = x} \left(\nu_1(y_1) \wedge \nu_2(y_2) \right).$$

Since ν_1 is increasing on the interval $[a_1, b_1]$ and ν_2 is increasing on the interval $[a_2, b_2]$, the vector (y_1, y_2) which maximizes the binary function $\nu_1 \wedge \nu_2$ should satisfy the following equations

$$\alpha_1 y_1 + \alpha_2 y_2 = x, \qquad (y_1 - a_1)/(b_1 - a_1) = (y_2 - a_2)/(b_2 - a_2).$$

It is solved that

$$y_2 = \frac{x(b_2 - a_2) + \alpha_1((b_1 - a_1)a_2 - a_1(b_2 - a_2))}{\alpha_1(b_1 - a_1) + \alpha_2(b_2 - a_2)}.$$

Taking it into the credibility function ν_2, we get

$$\nu(x) = (x - a)/2(b - a).$$

Case 3. $b \leq x < c$. It follows from the Zadeh extension theorem that

$$\nu(x) = \sup_{b_1 \leq y_1 \leq c_1, b_2 \leq y_2 \leq c_2, \alpha_1 y_1 + \alpha_2 y_2 = x} \left(\nu_1(y_1) \wedge \nu_2(y_2) \right).$$

Since ν_1 is decreasing on the interval $[b_1, c_1]$ and ν_2 is decreasing on the interval $[b_2, c_2]$, the vector (y_1, y_2) which maximizes the binary function $\nu_1 \wedge \nu_2$ should satisfy the following equations

$$\alpha_1 y_1 + \alpha_2 y_2 = x, \qquad (c_1 - y_1)/(c_1 - b_1) = (c_2 - y_2)/(c_2 - b_2).$$

It is solved that

$$y_2 = \frac{x(c_2 - b_2) + \alpha_1((c_1 - b_1)c_2 - c_1(c_2 - b_2))}{\alpha_1(c_1 - b_1) + \alpha_2(c_2 - b_2)}.$$

Taking it into the credibility function v_2, we get

$$v(x) = (c - x)/2(c - b).$$

Case 4. $x \geq c$. For any $\alpha_1 x_1 + \alpha_2 x_2 = x$, we have $x_1 \geq c_1$ or $x_2 \geq c_2$. It follows from the Zadeh extension theorem that $v(x) = 0$.

Remark 1.13 Similarly, suppose that trapezoidal fuzzy variables $\xi_1 = (a_1, b_1, c_1, d_1)$ and $\xi_2 = (a_2, b_2, c_2, d_2)$ are independent. Then for any nonnegative real numbers α_1 and α_2, we have

$$\alpha_1 \xi_1 + \alpha_2 \xi_2 = (\alpha_1 a_1 + \alpha_2 a_2, \alpha_1 b_1 + \alpha_2 b_2, \alpha_1 c_1 + \alpha_2 c_2, \alpha_1 d_1 + \alpha_2 d_2).$$

Example 1.24 Suppose that exponential fuzzy variables $\xi_1 = EXP(m_1)$ and $\xi_2 = EXP(m_2)$ are independent. Then for any $\alpha_1 > 0$ and $\alpha_2 > 0$, we have

$$\alpha_1 \xi_1 + \alpha_2 \xi_2 = E(\alpha_1 m_1 + \alpha_2 m_2). \tag{1.32}$$

Let v_1, v_2, and v be the credibility functions of fuzzy variables ξ_1, ξ_2, and $a_1 \xi_1 + a_2 \xi_2$, respectively. Since both v_1 and v_2 are strictly decreasing, for any $x > 0$, the vector (y_1, y_2) maximizing the binary function $v_1 \wedge v_2$ under the constraint $\alpha_1 y_1 + \alpha_2 y_2 = x$ should satisfy

$$v_1(y_1) = v_2(y_2).$$

It is solved that $y_1 = m_1 x/(a_1 m_1 + a_2 m_2)$. Taking it into the credibility function v_1, we get

$$v(x) = 1/\big(1 + \exp(\pi x/\sqrt{6}(a_1 m_1 + a_2 m_2))\big)$$

which implies that $a_1 \xi_1 + a_2 \xi_2$ is an exponential fuzzy variable with parameter $a_1 m_1 + a_2 m_2$.

Example 1.25 Suppose that normal fuzzy variables $\xi_1 = N(e_1, \sigma_1)$ and $\xi_2 = N(e_2, \sigma_2)$ are independent. Then for any $\alpha_1, \alpha_2 \in \Re$, we have

$$\alpha_1 \xi_1 + \alpha_2 \xi_2 = N\big(\alpha_1 e_1 + \alpha_2 e_2, |\alpha_1|\sigma_1 + |\alpha_2|\sigma_2\big). \tag{1.33}$$

Let v_1, v_2, and v be the credibility functions of fuzzy variables ξ_1, ξ_2, and $a_1 \xi_1 + a_2 \xi_2$, respectively. For simplicity, we denote $e = \alpha_1 e_1 + \alpha_2 e_2$ and $\sigma = |\alpha_1|\sigma_1 + |\alpha_2|\sigma_2$. In what follows, we will prove that

$$v(x) = 1/\big(1 + \exp(\pi|x - e|/\sqrt{6}\sigma)\big), \quad x \in \Re.$$

First, we assume $\alpha_1 \geq 0$ and $\alpha_2 \geq 0$. If $x > e$, since v_1 is strictly decreasing on the interval $[e_1, +\infty)$ and v_2 is strictly decreasing on the interval $[e_2, +\infty)$, the vector (y_1, y_2) maximizing the binary function $v_1 \wedge v_2$ should satisfy the following equations

$$\alpha_1 y_1 + \alpha_2 y_2 = x, \qquad (y_1 - e_1)/\sigma_1 = (y_2 - e_2)/\sigma_2,$$

which implies that

$$y_2 = \frac{x\sigma_2 + \alpha_1\sigma_1 e_2 - \alpha_1\sigma_2 e_1}{\alpha_1\sigma_1 + \alpha_2\sigma_2}.$$

Taking it into the credibility function v_2, we get

$$v(x) = 1/\left(1 + \exp\left(\pi(x - e)/\sqrt{6}\sigma\right)\right).$$

On the other hand, if $x \leq e$, since v_1 is strictly increasing on the interval $(-\infty, e_1]$ and v_2 is strictly increasing on the interval $(-\infty, e_2]$, the vector which maximizes the binary function $v_1 \wedge v_2$ should satisfy

$$\alpha_1 y_1 + \alpha_2 y_2 = x, \qquad (e_1 - y_1)/\sigma_1 = (e_2 - y_2)/\sigma_2,$$

which implies that

$$y_2 = \frac{x\sigma_2 + \alpha_1\sigma_1 e_2 - \alpha_1\sigma_2 e_1}{\alpha_1\sigma_1 + \alpha_2\sigma_2}.$$

Taking it into the credibility function v_2, we get

$$v(x) = 1/\left(1 + \exp\left(\pi(e - x)/\sqrt{6}\sigma\right)\right).$$

Let μ_1 be the credibility function of fuzzy variable $-\xi_1$. For any $x \in \Re$, it follows from the Zadeh extension theorem that

$$\mu_1(x) = v_1(-x) = 1/\left(1 + \exp\left(\pi|x + e_1|/\sqrt{6}\sigma_1\right)\right).$$

Therefore, for any $\alpha_1, \alpha_2 \in \Re$, we have

$$v(x) = 1/\left(1 + \exp\left(\pi|x - e|/\sqrt{6}\sigma\right)\right), \quad x \in \Re.$$

1.6 Identical Distribution

This section introduces the concept of identical distribution for fuzzy variables.

Definition 1.13 Suppose that fuzzy variables ξ and η have credibility functions μ and v, respectively. Then ξ and η are said to be identically distributed if and only if

$$v(x) = \mu(x), \quad \forall x \in \Re. \tag{1.34}$$

Theorem 1.21 *If fuzzy variables ξ and η are identically distributed, then for any set B of real numbers, we have*

$$\mathrm{Cr}\{\xi \in B\} = \mathrm{Cr}\{\eta \in B\}. \tag{1.35}$$

Proof It follows immediately from the credibility inversion theorem.   □

Theorem 1.22 *If fuzzy variables ξ and η are identically distributed, then for any real function f, fuzzy variables $f(\xi)$ and $f(\eta)$ are also identically distributed.*

Proof It follows immediately from the Zadeh extension theorem.   □

Example 1.26 If fuzzy variables ξ and η are identically distributed, then it follows from above theorem that ξ^2 and η^2 are also identically distributed.

Remark 1.14 Suppose that fuzzy variables ξ and η have credibility functions ν and μ, respectively. If $\xi = \eta$, then for any $x \in \Re$, we have

$$\nu(x) = \mathrm{Cr}\big\{\theta \in \Theta \mid \xi(\theta) = x\big\} = \mathrm{Cr}\big\{\theta \in \Theta \mid \eta(\theta) = x\big\} = \mu(x),$$

which implies that ξ and η are identically distributed. However, the inverse may be not true. For example, take a credibility space $(\Theta, \mathcal{A}, \mathrm{Cr})$ to be $\{\theta_1, \theta_2\}$ with $\mathrm{Cr}\{\theta_1\} = \mathrm{Cr}\{\theta_2\} = 0.5$. Define fuzzy variables

$$\xi = \begin{cases} 1, & \text{if } \theta = \theta_1 \\ -1, & \text{if } \theta = \theta_2, \end{cases} \qquad \eta = \begin{cases} -1, & \text{if } \theta = \theta_1 \\ 1, & \text{if } \theta = \theta_2. \end{cases}$$

It is clear that $\xi \neq \eta$, but they have the same credibility function

$$\nu(x) = \mu(x) = \begin{cases} 0.5, & \text{if } x \in \{-1, 1\} \\ 0, & \text{otherwise.} \end{cases}$$

Remark 1.15 Fuzzy variables $\xi_1, \xi_2, \ldots, \xi_m$ are said to be identically distributed if and only if each pair of them are identically distributed.

Example 1.27 Suppose that $\xi_i = (a, b, c), i = 1, 2, \ldots, m$ are independent and identically distributed triangular fuzzy variables. For any positive numbers $x_1, x_2, \ldots, x_m$ with $x_1 + x_2 + \cdots + x_m = 1$, we have

$$\xi_1 x_1 + \xi_2 x_2 + \cdots + \xi_m x_m = (a, b, c).$$

That is, the weighted sum is identically distributed with each element. The conclusion still holds if the triangular fuzzy variables are changed to be equipossible fuzzy variables, trapezoidal fuzzy variables, normal fuzzy variables, or exponential fuzzy variables.

References

De Cooman G (1997) Possibility theory I–III. Int J Gen Syst 25:291–371

Hisdal E (1978) Conditional possibilities independence and noninteraction. Fuzzy Sets Syst 1:283–297

Klir GJ (1999) On fuzzy-set interpretation of possibility theory. Fuzzy Sets Syst 108:263–273

Li X, Liu B (2006a) A sufficient and necessary condition for credibility measures. Int J Uncertain Fuzziness Knowl-Based Syst 14(5):527–535

Li X, Liu B (2006b) The independence of fuzzy variables with applications. Int J Nat Sci Technol 1(1):95–100

Li X, Liu B (2007) Maximum entropy principle for fuzzy variables. Int J Uncertain Fuzziness Knowl-Based Syst 15(2):43–52

Li PK, Liu B (2008a) Entropy of credibility distributions for fuzzy variables. IEEE Trans Fuzzy Syst 16(1):123–129

Li X, Liu B (2008b) On distance between fuzzy variables. J Intell Fuzzy Syst 19(3):197–204

Li X, Qin ZF, Kar S (2010a) Mean-variance-skewness model for portfolio selection with fuzzy returns. Eur J Oper Res 202:239–247

Liu B (2004) Uncertainty theory: an introduction to its axiomatic foundations. Springer, Berlin

Liu YK, Gao J (2007) The independence of fuzzy variables with applications to fuzzy random optimization. Int J Uncertain Fuzziness Knowl-Based Syst 15(2):1–20

Liu B, Liu YK (2002) Expected value of fuzzy variable and fuzzy expected value models. IEEE Trans Fuzzy Syst 10(4):445–450

Qin ZF, Li X, Ji XY (2009) Portfolio selection based on fuzzy cross-entropy. J Comput Appl Math 228:139–149

Yager RR (1992) On the specificity of a possibility distribution. J Cybern 50:279–292

Zadeh LA (1965) Fuzzy sets. Inf Control 8:338–353

Zadeh LA (1975) The concept of a linguistic variable and its application to approximate reasoning. Inf Sci 8:199–249

Zadeh LA (1978) Fuzzy sets as a basis for a theory of possibility. Fuzzy Sets Syst 1:3–28

Zadeh LA (1979) A theory of approximate reasoning. In: Hayes J, Michie D, Thrall RM (eds) Mathematical frontiers of the social and policy sciences. Westview Press, Boulder, pp 69–129

Chapter 2
Credibilistic Programming

The decision analysis with fuzzy objective or fuzzy constraints is natural in some real-world applications, and sometimes such analysis seems to be inevitable. Credibilistic programming is a type of mathematical programming for handling the fuzzy decision problems. In the past years, researchers have proposed various efficient modeling approaches based on different fuzzy ranking criteria. For example, Liu and Liu (2002) introduced a concept of expected value operator and then provided a spectrum of expected value model to maximize the average objective under certain expected constraints. Liu and Iwamura (1998a,b) introduced a maximax chance-constrained programming model, and Liu (1998) provided a maximin chance-constrained programming model, which respectively maximizes the optimistic objective and pessimistic objective under certain credibility constraints. Based on the concepts of fuzzy entropy, Li et al. (2011) formulated an entropy optimization model, which was extended by Qin et al. (2009) to the cross-entropy minimization model. Recently, Li et al. (2012) introduced a regret minimization model to minimize the distance between the fuzzy objective values and the best values.

This chapter mainly provides a general description on nonlinear programming, multi-objective programming, and credibilistic programming. In addition, a brief introduction on the solution methods will also be given, including the Kuhn-Tucker conditions and genetic algorithm.

2.1 Mathematical Programming

As one of the most widely used technique in operations research, mathematical programming is defined as a means of maximizing a quantity known as objective function, subject to a set of constraints. It is impossible that this section covers all concepts of mathematical programming. Therefore, this section only introduces some basic concepts and techniques such that readers can gain an understanding of them throughout the book.

X. Li, *Credibilistic Programming*, Uncertainty and Operations Research,
DOI 10.1007/978-3-642-36376-4_2, © Springer-Verlag Berlin Heidelberg 2013

2.1.1 Single-Objective Programming

In mathematical terms, the general form of a single-objective programming can be written as follows:

$$\begin{cases} \max & f(x) \\ \text{s.t.} & g_i(x) \le 0, \quad i = 1, 2, \ldots, n \end{cases} \tag{2.1}$$

where $x = (x_1, x_2, \ldots, x_m)$ is the decision vector, the first line defines the objective function to be maximized, and the second line defines the inequality constraints.

Definition 2.1 For the single-objective programming model (2.1), the set

$$S = \{x \in \Re^m \mid g_i(x) \le 0, \ i = 1, 2, \ldots, n\} \tag{2.2}$$

is called the feasible set. An element x in S is called a feasible solution.

Definition 2.2 For the single-objective programming model (2.1), a feasible solution x^* is called the local optimal solution if and only if there is a real number $\varepsilon > 0$ such that

$$f(x^*) \ge f(x) \tag{2.3}$$

for all feasible solution x with $\|x - x^*\| < \varepsilon$.

Definition 2.3 For the single-objective programming model (2.1), a feasible solution x^* is called the global optimal solution if and only if

$$f(x^*) \ge f(x) \tag{2.4}$$

for all feasible solution $x \in S$.

Remark 2.1 Note that a global optimal solution must be a local optimal solution, but a local optimal solution may be not a global optimal solution.

Example 2.1 In order to illustrate the concepts of feasible solution, local optimal solution, and global optimal solution, we consider the following single-objective programming problem

$$\begin{cases} \max & \max\{(x-1)(2-x), 1 - x^2, -(x+1)(x+2)\} \\ \text{s.t.} & (x+2)(x-2) \le 0. \end{cases}$$

It is easy to prove that the feasible set is a closed interval $S = [-2, 2]$. There are three local optimal solutions $x_1 = -1.5$, $x_2 = 0$, $x_3 = 1.5$, among which $x_2 = 0$ is the global optimal solution. See Fig. 2.1.

Fig. 2.1 Local optimal
solution and global optimal
solution

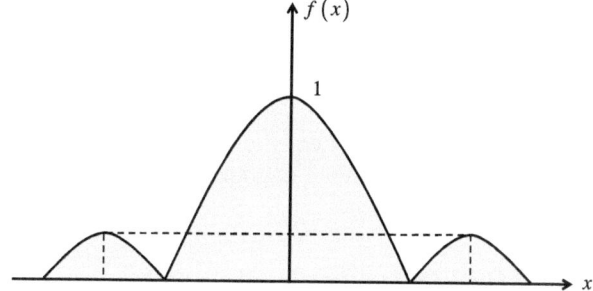

One of the most outstanding contributions to mathematical programming is
known as the *Kuhn-Tucker conditions*. In order to introduce them, we first give
some definitions. An inequality constraint $g_i \leq 0$ is said to be *active* at a point x
if $g_i(x) = 0$. A feasible point x is said to be *regular* if the gradient vector $\nabla g_i(x)$
of all active constraints are linearly independent.

Suppose that x^* is a regular point of the single-objective programming model
(2.1), and all the functions f and g_i, $i = 1, 2, \ldots, n$ are differentiable. If x^* is a
local optimal solution, then there exist Lagrangian multipliers λ_i, $i = 1, 2, \ldots, n$
such that the following Kuhn-Tucker conditions hold,

$$\begin{cases} \nabla f(x^*) - \sum_{i=1}^{n} \lambda_i g_i(x^*) = 0 \\ \lambda_i g_i(x^*) = 0, \quad i = 1, 2, \ldots, n \\ \lambda_i \geq 0, \quad i = 1, 2, \ldots, n. \end{cases} \tag{2.5}$$

Furthermore, if functions g_i, $i = 1, 2, \ldots, n$ are all convex and the objective function
f is concave, then a regular point x^* is the global optimal solution if and only if it
satisfies the Kuhn-Tucker conditions.

Example 2.2 In this example, we apply the Kuhn-Tucker conditions to solve the
following single-objective programming problem

$$\begin{cases} \max & -x_1^2 - x_2^2 - x_3^2 \\ \text{s.t.} & 1 - x_1 - x_2 - x_3 \leq 0 \\ & x_1, x_2, x_3 \geq 0. \end{cases}$$

It is clear that the objective function $f = -x_1^2 - x_2^2 - x_3^2$ is differentiable and con-
cave, and the constraint function $g = 1 - x_1 - x_2 - x_3$ is differentiable and convex.
Therefore, a point is the global optimal solution if and only if it satisfies the Kuhn-
Tucker conditions

$$\begin{cases} -2x_i + \lambda = 0, \quad i = 1, 2, 3 \\ \lambda(1 - x_1 - x_2 - x_3) = 0 \\ \lambda \geq 0 \end{cases}$$

where λ is the Lagrangian multiplier. According to the inequality constraint, there is at least one index i such that $x_i > 0$, which implies that $\lambda > 0$ and

$$x_1 + x_2 + x_3 = 1.$$

Taking $x_i = \lambda/2$ into this equation, it is solved that the global optimal solution is $x_1 = x_2 = x_3 = 1/3$, and the Lagrangian multiplier is $\lambda = 2/3$.

2.1.2 Multi-Objective Programming

In multi-objective programming, also known as multi-attribute programming or multi-criteria programming, we attempt to simultaneously maximize two or more conflicting objectives subject to certain constraints, which is formulated as follows,

$$\begin{cases} \max & [f_1(x), f_2(x), \ldots, f_p(x)] \\ \text{s.t.} & g_i(x) \le 0, \quad i = 1, 2, \ldots, n. \end{cases} \tag{2.6}$$

Similarly, the set $S = \{x \in \Re^m \mid g_i(x) \le 0, \ i = 1, 2, \ldots, n\}$ is called the feasible set, and each element x of S is called a feasible solution.

For a nontrivial multi-objective programming problem, one cannot identify a solution that simultaneously maximizes all objectives. If the decision-maker has a real preference function which aggregates all the objectives, then we may maximize the preference function under the same set of constraints. The obtained single-objective programming model is called a *compromise model* whose solution is called a *compromise solution*.

The first well-known compromise model is formulated to maximize the linearly weighted objective function

$$\begin{cases} \max & \sum_{i=1}^{p} \lambda_i f_i(x) \\ \text{s.t.} & g_i(x) \le 0, \quad i = 1, 2, \ldots, n \end{cases} \tag{2.7}$$

where $\lambda_1, \lambda_2, \ldots, \lambda_p$ are nonnegative real numbers, which denote the preferences of the decision-maker on different objectives. Taking a two-objective programming model for example, if the first objective is more important than the second one, we set $\lambda_1 > \lambda_2$. Otherwise, we set $\lambda_1 \le \lambda_2$.

The second way is formulated to minimize the distance between the objective vector and an ideal vector $(f_1^*, f_2^*, \ldots, f_p^*)$, where f_i^* is the maximum value for the ith objective without considering other objectives. If the Euclidean distance is used, we have

$$\begin{cases} \min & \sqrt{\sum_{i=1}^{p} \left(f_i(x) - f_i^* \right)^2} \\ \text{s.t.} & g_i(x) \le 0, \quad i = 1, 2, \ldots, n. \end{cases} \tag{2.8}$$

Table 2.1 Database for notebook computer selection

Candidate	Price (RMB)	Weight (kg)	Size (inch)
x_1	6730	2.2	15.4
x_2	6980	2.2	15.4
x_3	9200	1.3	12.1
x_4	13000	2.2	12.1

The term of Pareto optimality and the related terms of Pareto dominance, Pareto solution, and Pareto set are the most basic concepts in multi-objective programming theory (Ehrgott 2000; Farina and Amato 2004; Li and Wong 2009; Pierro et al. 2007) and algorithms (Chou et al. 2008; Delgado et al. 2008; Ewald et al. 2008; Hung et al. 2008; Tan et al. 2005; Zou et al. 2008). Roughly speaking, Pareto optimality means that when we attempt to improve an objective further, other objectives suffer as a result.

Definition 2.4 For any feasible solutions x, $y \in S$, x is said to Pareto dominate y if and only if

(a) $f_i(x) \geq f_i(y)$ for all $i \in \{1, 2, \ldots, p\}$;
(b) $f_j(x) > f_j(y)$ for at least one index $j \in \{1, 2, \ldots, p\}$.

Definition 2.5 A feasible solution x is said to be a Pareto solution if there is no feasible solution y which Pareto dominates x. The set of all Pareto solutions is called the Pareto set.

Example 2.3 Suppose that we would like to select a cheap, light and small notebook computer from the candidates $\{x_1, x_2, x_3, x_4\}$. The detailed data about each candidate is shown in Table 2.1. It is easy to prove that x_1 and x_3 are Pareto solutions. Note that x_2 is not a Pareto solution since it is dominated by x_1, and x_4 is not a Pareto solution since it is dominated by x_3.

Remark 2.2 The global optimal solution x^* of compromise model (2.7) must be a Pareto solution. Otherwise, according to Definition 2.5, there is a feasible solution x such that $f_i(x) \geq f_i(x^*)$ for all $i = 1, 2, \ldots, p$, and the strict inequality holds for at least one index. Then it is easy to prove that

$$\sum_{i=1}^{p} \lambda_i f_i(x) > \sum_{i=1}^{p} \lambda_i f_i(x^*),$$

which is in contradiction with the fact that x^* is the global optimal solution. Similarly, we can prove that the global optimal solution of compromise model (2.8) is also a Pareto solution.

2.2 Credibilistic Programming

Fuzzy programming is the mathematical programming in fuzzy environment, that is, the objective function f or constraint functions $g_i, i = 1, 2, \ldots, n$ contain fuzzy parameters. Assume that x is a decision vector, and ξ is a fuzzy vector, then the general fuzzy programming model can be written as

$$\begin{cases} \max & f(x, \xi) \\ \text{s.t.} & g_i(x, \xi) \leq 0, \quad i = 1, 2, \ldots, n. \end{cases} \tag{2.9}$$

Example 2.4 In this example, we consider the portfolio selection problem. The term portfolio refers to any collection of financial assets such as stocks, bonds, and cash. Portfolio may be held by individual investors or managed by financial professionals, banks and other financial institutions.

Assume that there are m stocks, and we use ξ_i to denote the return of the ith stock. In general, ξ_i is given as $(p_i' + d_i - p_i)/p_i$ where p_i is the closing price at present, p_i' is the closing price in the next year, and d_i is the dividend during the coming year. Note that the values of p_i' and d_i in a future time period are clearly unknown at present. If they are estimated as fuzzy quantities, then ξ_i is a fuzzy variable. Furthermore, for each portfolio $(x_1, x_2, \ldots, x_m)$, where x_i denotes the proportion of the total capital invested in stock i, the total return

$$f(x, \xi) = \xi_1 x_1 + \xi_2 x_2 + \cdots + \xi_m x_m$$

is also a fuzzy variable. In this case, if the investor would like to maximize the total return, we get the following fuzzy programming model

$$\begin{cases} \max & \xi_1 x_1 + \xi_2 x_2 + \cdots + \xi_m x_m \\ \text{s.t.} & x_1 + x_2 + \cdots + x_m = 1 \\ & x_i \geq 0, \quad i = 1, 2, \ldots, m \end{cases} \tag{2.10}$$

where the first constraint implies that all the capital will be invested to the m stocks, and the next set of constraints implies that short sale and borrowing are not allowed.

Generally speaking, it is meaningless to maximize a fuzzy objective since there is not a natural ordership in fuzzy world. Therefore, we need to define a *credibilistic mapping* from the collection of fuzzy variables to the set of real numbers, such that we can rank fuzzy variables according to the natural ordership of real numbers. For the fuzzy programming model (2.9), if the credibilistic mappings $U, U_1, U_2, \ldots, U_n$ are taken, we get the following model

$$\begin{cases} \max & U\big[f(x, \xi)\big] \\ \text{s.t.} & U_i\big[g_i(x, \xi)\big] \leq 0, \quad i = 1, 2, \ldots, n. \end{cases} \tag{2.11}$$

Note that (2.11) is a crisp nonlinear programming model since the objective function and constraints are both well defined. In what follows, we will call it a *credibilistic programming* model. The following chapters will introduce some mainly used credibilistic mappings including the expected value operator, optimistic value, pessimistic value, entropy, cross-entropy, and distance.

Definition 2.6 For the credibilistic programming model (2.11), the set

$$S = \left\{ x \in \Re^m \mid U_i\big[g_i(x, \xi)\big] \leq 0, \ i = 1, 2, \ldots, n \right\} \tag{2.12}$$

is called the feasible set. An element x in S is called a feasible solution.

Definition 2.7 For the credibilistic programming model (2.11), a feasible solution x^* is called the local optimal solution if and only if there is a real number $\varepsilon > 0$ such that

$$U\big[f(x^*, \xi)\big] \geq U\big[f(x, \xi)\big] \tag{2.13}$$

for all feasible solution x with $\|x - x^*\| < \varepsilon$.

Definition 2.8 For the credibilistic programming model (2.11), a feasible solution x^* is called the global optimal solution if and only if

$$U\big[f(x^*, \xi)\big] \geq U\big[f(x, \xi)\big] \tag{2.14}$$

for all feasible solution $x \in S$.

If there are multiple objective functions $f_1, f_2, \ldots, f_p$, we can define the following multi-objective credibilistic programming model,

$$\begin{cases} \max & \big[U\big[f_1(x, \xi)\big], U\big[f_2(x, \xi)\big], \ldots, U\big[f_p(x, \xi)\big]\big] \\ \text{s.t.} & U_i\big[g_i(x, \xi)\big] \leq 0, \quad i = 1, 2, \ldots, n. \end{cases} \tag{2.15}$$

Definition 2.9 For any feasible solutions $x, y \in S$, x is said to Pareto dominate y if and only if

(a) $U[f_i(x, \xi)] \geq U[f_i(y, \xi)]$ for all $i \in \{1, 2, \ldots, m\}$;
(b) $U[f_j(x, \xi)] > U[f_j(y, \xi)]$ for at least one index $j \in \{1, 2, \ldots, m\}$.

Definition 2.10 A feasible solution x is said to be a Pareto solution for the multi-objective credibilistic programming model (2.15) if there is no feasible solution y which Pareto dominates x. The set of all Pareto solutions is called the Pareto set.

Example 2.5 Let us reconsider the portfolio selection problem. Suppose that there are three stocks and the returns are independent triangular fuzzy variables $\xi_1 = (0, 3, 6)$, $\xi_2 = (2, 3, 4)$, and $\xi_3 = (-1, 0, 1)$. See Fig. 2.2.

It is clear that the second stock has a better return than the third stock since it follows from the credibility inversion theorem that $\text{Cr}\{\xi_2 \geq \xi_3\} = 1$. However, it is

Fig. 2.2 Credibility
functions for ξ_1, ξ_2 and ξ_3

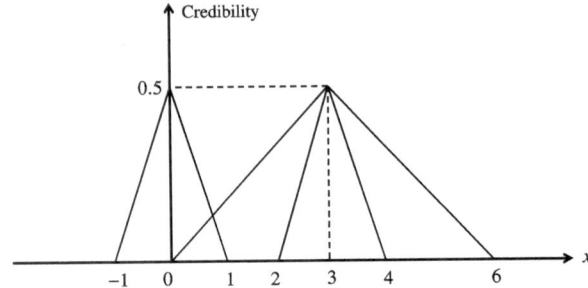

difficult to compare the returns arising from the second stock and the first stock.
As a result, if we take portfolios $x = (1,0,0)$ and $y = (0,1,0)$, it is difficult to
decide which one is better. In fact, it follows form the independence that fuzzy
vector (ξ_1,ξ_2) has a joint credibility function $v = v_1 \wedge v_2$. Then according to the
credibility inversion theorem, event $\{\xi_1 \geq \xi_2\}$ has a credibility

$$\mathrm{Cr}\{\xi_1 \geq \xi_2\} = 1 - \sup_{x_1 < x_2} v(x_1, x_2)$$

$$= 1 - \sup_{x_1 < 3, x_2 = 3} \big(v_1(x_1) \wedge v_2(x_2)\big)$$

$$= 0.5,$$

and event $\{\xi_2 \geq \xi_1\}$ has a credibility

$$\mathrm{Cr}\{\xi_2 \geq \xi_1\} = 1 - \sup_{x_2 < 3, x_1 = 3} \big(v_1(x_1) \wedge v_2(x_2)\big) = 0.5.$$

If there is a credibilistic mapping such that $U(\xi) = a$ for each triangular fuzzy
variable $\xi = (a, b, c)$, then the credibilistic programming model is

$$\begin{cases} \max & 2x_2 - x_3 \\ \text{s.t.} & x_1 + x_2 + x_3 = 1 \\ & x_1, x_2, x_3 \geq 0. \end{cases}$$

It is easy to calculate that the optimal portfolio is $x^* = (0,1,0)$. If there is another
credibilistic mapping which takes value c for each triangular fuzzy variable $\xi =
(a, b, c)$, we have the following credibilistic programming model

$$\begin{cases} \max & 6x_1 + 4x_2 + x_3 \\ \text{s.t.} & x_1 + x_2 + x_3 = 1 \\ & x_1, x_2, x_3 \geq 0. \end{cases}$$

In this case, the optimal portfolio is $x^* = (1,0,0)$.

2.3 Genetic Algorithm

For a general credibilistic programming model, if the credibilistic mappings have analytical expressions and the objective and constraint functions have good mathematical properties, such as differentiability and convexity, we can design efficient solution algorithms by using the Kuhn-Tucker conditions. However, the credibilistic mappings generally have no analytical expressions or the expressions have bad properties. In this case, we can try to obtain suboptimal solution by using the genetic algorithm.

Genetic algorithm is a stochastic search method for optimization problems based on the mechanics of natural selection and natural genetics, i.e., survival of the fittest, which has been well-documented in the literatures, such as in Holland (1975), Goldberg (1989), Michalewicz (1996), Koza (1992, 1994), and so on. In the past decades, genetic algorithm has obtained considerable success in providing satisfactory solutions to many complex optimization problems and received more and more attentions. This section introduces the basic steps for genetic algorithm including representation structure, initialization, evaluation function, selection process, crossover operation, and mutation operation. At the end of this section, a general procedure of the genetic algorithm is also given.

2.3.1 Representation Structure

The first problem for genetic algorithm is how to construct a one to one mapping between the solution space and the chromosome space such that the following operations such as crossover and mutation, can be simplified. The mapping from the solution space to the chromosome space is called *encoding*, and the mapping from the chromosome space to the solution space is called *decoding*.

The representation of solution is generally problem dependent, while binary encoding and floating encoding are two mainly used representation structures. Taking the floating encoding for example, let $x = (x_1, x_2, \ldots, x_m)$ be a solution vector in the solution space satisfying

$$\begin{cases} x_1 + x_2 + \cdots + x_m = 1 \\ x_i \geq 0, \quad i = 1, 2, \ldots, m. \end{cases} \tag{2.16}$$

We may encode the solution by a chromosome $\mathbf{v} = (v_1, v_2, \ldots, v_m)$ satisfying

$$v_i \geq 0, \quad i = 1, 2, \ldots, m. \tag{2.17}$$

Then the encoding and decoding processes are determined by the equations

$$x_1 = \frac{v_1}{v_1 + v_2 + \cdots + v_m}, \quad i = 1, 2, \ldots, m. \tag{2.18}$$

2.3.2 Initialization

Define an integer *pop-size* as the size of population, which generally depends on the nature of the problem. Randomly generate *pop-size* chromosomes for the initialized population. Usually, it is difficult to produce feasible chromosomes explicitly for complex optimization problems with irregular feasible set. However, if the decision-maker can predetermine a region with regular sharp which contains the optimal solution, we may initialize the population in the regular field.

We generate a random point from the region and check its feasibility. If it is feasible, then it will be accepted as a chromosome. If not, we regenerate a point randomly until a feasible one is obtained. We repeat this procedure *pop-size* times, and generate the first population. The initialization process is summarized as follows.

Algorithm 2.1 (Initialization Process)

Step 1. Set $i = 1$.
Step 2. Randomly generate a chromosome from the predetermined region.
Step 3. If it is feasible, set $i = i + 1$. Otherwise, go to step 2.
Step 4. If $i \leq pop\text{-}size$, go to step 2.
Step 5. Return the initialized population $\mathbf{v}_i$, $i = 1, 2, \ldots, pop\text{-}size$.

2.3.3 Evaluation Function

Evaluation function assigns each chromosome a probability of reproduction so that its likelihood of being selected is proportional to its fitness relative to the other chromosomes in the population. That is, the chromosomes with higher fitness will have more chance to produce offspring.

Assume that the decision-maker can give an order relationship among these *pop-size* chromosomes such that they are rearranged from good to bad. For example, for a single-objective programming problem, a chromosome with a larger objective value is better, while for a multi-objective programming problem, a chromosome with a larger aggregating preference function is better. One well-known evaluation function is based on allocation of reproductive trials according to rank rather than actual objective values. For each $\alpha \in (0, 1)$, we define the rank-based evaluation function as follows,

$$Eval(\mathbf{v}_i) = \alpha(1 - \alpha)^{i-1}, \quad i = 1, 2, \ldots, pop\text{-}size. \tag{2.19}$$

Note that $\mathbf{v}_1$ is the best chromosome, and $\mathbf{v}_{pop\text{-}size}$ is the worst one.

Algorithm 2.2 (Evaluation Process)

Step 1. Initialize a real number $\alpha \in (0, 1)$.
Step 2. Calculate the objective values f_i for all chromosomes.

Step 3. Reorder these chromosomes according to their objective values.
Step 4. Set $i = 1$.
Step 5. Calculate the evaluation value for the ith chromosome

$$Eval(\mathbf{v}_i) = \alpha(1 - \alpha)^{i-1}.$$

Step 6. If $i < pop\text{-}size$, set $i = i + 1$, and goto step 5.

2.3.4 Selection Process

During each successive generation, a proportion of the existing population is selected to breed a new generation. The selection process is based on spinning the roulette wheel $pop\text{-}size$ times, and selecting a single chromosome at each time. The roulette wheel is a fitness-proportional selection, where fitter chromosomes (as measured by the objective value) are typically more likely to be selected. The selection process is summarized as follows.

Algorithm 2.3 (Selection Process)

Step 1. Calculate the reproduction probability q_i for each chromosome $\mathbf{v}_i$,

$$q_0 = 0, \quad q_i = \sum_{j=1}^{i} Eval(\mathbf{v}_j), \quad i = 1, 2, \ldots, pop\text{-}size.$$

Step 2. Generate a random number r in $(0, q_{pop\text{-}size}]$.
Step 3. Select the chromosome $\mathbf{v}_i$ such that $q_{i-1} < r \le q_i$.
Step 4. Repeat the second and third steps $pop\text{-}size$ times and obtain $pop\text{-}size$ chromosomes.

2.3.5 Crossover Operation

Crossover is one of the mainly used operations for generating a second population. First, we define a parameter P_c to denote the probability of crossover. Then we randomly select some chromosomes as parents from the pool selected previously. Repeat the following process $pop\text{-}size$ times: generate a random number r from $[0, 1]$, and select the chromosome $\mathbf{v}_i$ if $r < p_c$. Note that not all chromosomes can be selected, and different chromosomes have the equal chance to be selected. Denote the selected parents by $\mathbf{u}_1, \mathbf{u}_2, \mathbf{u}_3, \ldots, \mathbf{u}_c$ and divide them into the following pairs:

$$(\mathbf{u}_1, \mathbf{u}_2), \quad (\mathbf{u}_3, \mathbf{u}_4), \quad (\mathbf{u}_5, \mathbf{u}_6), \quad \ldots$$

Let us illustrate the crossover operation by the first pair $(\mathbf{u}_1, \mathbf{u}_2)$. Generate a random number λ from the open interval $(0, 1)$, then the crossover operator on $\mathbf{u}_1$ and $\mathbf{u}_2$ will produce two children x and y as follows:

$$\mathbf{x} = \lambda \mathbf{u}_1 + (1 - \lambda)\mathbf{u}_2, \qquad \mathbf{y} = (1 - \lambda)\mathbf{u}_1 + \lambda \mathbf{u}_2.$$

We check the feasibility for each child before accepting it. If both children are feasible, then we replace the parents with them. If not, we keep the feasible one if it exists, and then redo the crossover operator until two feasible children are obtained or a number of cycles is finished.

Algorithm 2.4 (Crossover Operation)

Step 1. Initialize a crossover probability P_c, and set $i = 1$.
Step 2. Generate a random number r from $[0, 1]$.
Step 3. If $r \leq P_c$, select chromosome $\mathbf{v}_i$ as the parent, and set $i = i + 1$.
Step 4. If $i \leq pop\text{-}size$, go to step 2.
Step 5. Denote the selected parents by $\mathbf{u}_1, \mathbf{u}_2, \mathbf{u}_3, \ldots, \mathbf{u}_c$, and set $j = 1$.
Step 6. Generate a random number λ from $(0, 1)$, and produce two children

$$\mathbf{x} = \lambda \mathbf{u}_j + (1 - \lambda)\mathbf{u}_{j+1}, \qquad \mathbf{y} = (1 - \lambda)\mathbf{u}_j + \lambda \mathbf{u}_{j+1}.$$

Redo this operation until two feasible children are obtained or a number of cycles is finished.
Step 7. Replace the parents with the feasible children, and set $j = j + 2$.
Step 8. If $j \leq c$, go to step 6.

2.3.6 Mutation Operation

Mutation is another operation for updating the chromosomes. We define a parameter P_m to denote the probability of mutation, and randomly select some chromosomes as parents in a similar way to the process of selecting parents for crossover.

For each selected parent $\mathbf{v}$, we mutate it in the following way. Let λ be an approximate large positive number, and let d be a mutation direction. If $\mathbf{v} + \lambda d$ is not feasible, then we randomly decrease the value of λ until it is feasible. If the above process cannot find a feasible solution in a predetermined number of iterations, we set $\lambda = 0$. Anyway, we replace the parent chromosome $\mathbf{v}$ with its child $\mathbf{v} + \lambda d$. The mutation operation is summarized as follows.

Algorithm 2.5 (Mutation Operation)

Step 1. Initialize a mutation probability P_m, and set $i = 1$.
Step 2. Generate a random number r from $[0, 1]$.
Step 3. If $r \leq P_m$, generate a mutation direction d and a parameter λ such that $\mathbf{v}_i + \lambda d$ is a feasible chromosome. Replace $\mathbf{v}_i$ with $\mathbf{v}_i + \lambda d$.
Step 4. Set $i = i + 1$.
Step 5. If $i \leq pop\text{-}size$, go to step 2.

2.3.7 General Procedure

Following selection, crossover and mutation operations, a new population is generated. Genetic algorithm will terminate after a given number of cyclic iterations of the above steps. We now summarize the general procedure for genetic algorithm as follows.

Algorithm 2.6 (Genetic Algorithm)

Step 1. Randomly Initialize *pop-size* chromosomes.
Step 2. Calculate the objective values for all chromosomes.
Step 3. Evaluate the fitness of each chromosome via the objective values.
Step 4. Select the chromosomes by spinning the roulette wheel.
Step 5. Update the chromosomes by using crossover and mutation.
Step 6. Repeat the second to fifth steps for a given number of cycles.
Step 7. Report the best found chromosome as the suboptimal solution.

References

Chou TY, Liu TK, Lee CN, Jeng CR (2008) Method of inequality-based multiobjective genetic algorithm for domestic daily aircraft routing. IEEE Trans Syst Man Cybern, Part A 38(2):299–308

Delgado M, Cuellar MP, Pegalajar MC (2008) Multiobjective hybrid optimization and training of recurrent neural networks. IEEE Trans Syst Man Cybern, Part B 38(2):381–403

Ehrgott M (2000) Multicriteria optimization. Springer, Berlin

Ewald G, Kurek W, Brdys MA (2008) Grid implementation of a parallel multiobjective genetic algorithm for optimized allocation of chlorination stations in drinking water distribution systems: Chojnice case study. IEEE Trans Syst Man Cybern, Part C 38(4):497–509

Farina M, Amato P (2004) A fuzzy definition of "optimality" for many-criteria optimization problems. IEEE Trans Syst Man Cybern, Part A 34(2):315–326

Goldberg DE (1989) Genetic algorithms & engineering optimization. Wiley, New York

Holland JH (1975) Adaptation in natural and artificial systems. University of Michigan Press, Ann Arbor

Hung MH, Shu LS, Ho SJ, Hwang SF, Ho SY (2008) A novel intelligent multiobjective simulated annealing algorithm for designing robust PID controllers. IEEE Trans Syst Man Cybern, Part A 38(2):319–330

Koza JR (1992) Genetic programming. MIT Press, Cambridge

Koza JR (1994) Genetic programming, II. MIT Press, Cambridge

Li X, Ralescu D, Tang T (2011) Fuzzy train energy consumption minimization model and algorithm. Iran J Fuzzy Syst 8(4):77–91

Li X, Shou BY, Qin ZF (2012) An expected regret minimization portfolio selection model. Eur J Oper Res 218:484–492

Li X, Wong HS (2009) Logic optimality for multi-objective optimization. Appl Math Comput 215:3045–3056

Liu B (1998) Minimax chance constrained programming models for fuzzy decision systems. Inf Sci 112(1–4):25–38

Liu B, Iwamura K (1998a) Chance constrained programming with fuzzy parameters. Fuzzy Sets Syst 94(2):227–237

Liu B, Iwamura K (1998b) A note on chance constrained programming with fuzzy coefficients. Fuzzy Sets Syst 100(1–3):229–233

Liu B, Liu YK (2002) Expected value of fuzzy variable and fuzzy expected value models. IEEE Trans Fuzzy Syst 10(4):445–450

Michalewicz Z (1996) Genetic algorithms + data structures = evolution programs, 3rd edn. Springer, Berlin

Pierro F, Khu ST, Savic DA (2007) An investigation on preference order ranking scheme for multiobjective evolutionary optimization. IEEE Trans Evol Comput 11(1):17–45

Qin ZF, Li X, Ji XY (2009) Portfolio selection based on fuzzy cross-entropy. J Comput Appl Math 228:139–149

Tan KC, Khor EF, Lee TH (2005) Multiobjective evolutionary algorithms and applications. Springer, London

Zou XF, Chen Y, Liu MZ, Kang JS (2008) A new evolutionary algorithm for solving many-objective optimization problems. IEEE Trans Syst Man Cybern, Part B 38(5):1402–1412

Chapter 3
Expected Value Model

Expected value (or mean value) of a fuzzy variable is the weighted average of all possible values in the sense of credibility measure, which is one of the most well-known credibilistic mappings for ranking fuzzy variables (see literatures (Dubois and Prade 1987; González 1990; Heilpern 1992; Liu and Liu 2002; Zhu and Ji 2006)). Based on the concept of expected value, Liu and Liu (2002) proposed an *expected value model*, which had been widely used in many real-life applications, such as newsboy problem (Ji and Shao 2006; Shao and Ji 2006), facility location problem (Zhou and Liu 2007), parallel machine scheduling problem (Peng and Liu 2004), portfolio selection problem (Li et al. 2010a), project scheduling problem (Ke and Liu 2010), risk assessment problem (Feng et al. 2008), shortest path problem (Ji and Iwamura 2007), system reliability design (Zhao and Liu 2005), supply chain design (Das et al. 2007), train scheduling problem (Yang et al. 2009), and so on.

This chapter mainly introduces the concepts of expected value, variance, skewness, moment, as well as the fuzzy simulation technique, expected value model and applications in fuzzy portfolio analysis.

3.1 Expected Value

This section introduces the concepts of expected value, variance, skewness, and moment. Some important theorems are also proved, including the expected value linearity theorem, maximum variance theorem, and so on.

Definition 3.1 (Liu and Liu 2002) Let ξ be a fuzzy variable on credibility space $(\Theta, \mathcal{A}, \mathrm{Cr})$. Then its expected value is defined by

$$E[\xi] = \int_0^{+\infty} \mathrm{Cr}\{\xi \geq r\}\,\mathrm{d}r - \int_{-\infty}^0 \mathrm{Cr}\{\xi \leq r\}\,\mathrm{d}r \qquad (3.1)$$

provided that at least one of the two integrals is finite.

X. Li, *Credibilistic Programming*, Uncertainty and Operations Research,
DOI 10.1007/978-3-642-36376-4_3, © Springer-Verlag Berlin Heidelberg 2013

Remark 3.1 Let ξ be a positive fuzzy variable. For any $r \le 0$, we have $\mathrm{Cr}\{\xi \le r\} = 0$. Then it follows from Definition 3.1 that

$$E[\xi] = \int_0^{+\infty} \mathrm{Cr}\{\xi \ge r\}\,\mathrm{d}r$$

which is clear a positive quantity.

Remark 3.2 Let ξ be a negative fuzzy variable. For any $r \ge 0$, we have $\mathrm{Cr}\{\xi \ge r\} = 0$. Then it follows from Definition 3.1 that

$$E[\xi] = -\int_{-\infty}^0 \mathrm{Cr}\{\xi \le r\}\,\mathrm{d}r$$

which is a negative quantity.

Remark 3.3 It is possible that the expected values for some fuzzy variables are infinite. For example, let ξ be a nonnegative fuzzy variable defined by the following credibility function

$$\nu(x) = \begin{cases} 1/(2+x), & \text{if } x \ge 0 \\ 0, & \text{if } x < 0. \end{cases}$$

According to the credibility inversion theorem, it is easy to prove that

$$\mathrm{Cr}\{\xi \ge r\} = 1/(2+r)$$

for all $r > 0$. Then it follows form Definition 3.1 that

$$E[\xi] = \int_0^\infty 1/(2+r)\,\mathrm{d}r = +\infty.$$

On the other hand, let ξ be a negative fuzzy variable with credibility function

$$\nu(x) = \begin{cases} 1/(2-x), & \text{if } x < 0 \\ 0, & \text{if } x \ge 0. \end{cases}$$

According to the credibility inversion theorem, it is easy to prove that

$$\mathrm{Cr}\{\xi \le r\} = 1/(2-r)$$

for all $r < 0$. Then it follows form Definition 3.1 that

$$E[\xi] = -\int_{-\infty}^0 1/(2-r)\,\mathrm{d}r = -\infty.$$

Remark 3.4 The expected value may not exist for some fuzzy variables. For example, a fuzzy variable defined by the following credibility function

$$\nu(x) = 1/(2+|x|), \quad x \in \Re$$

does not have expected value since

$$\int_0^\infty \mathrm{Cr}\{\xi \geq r\}\,dr = \int_0^\infty 1/(2+r)\,dr = +\infty,$$

$$\int_{-\infty}^0 \mathrm{Cr}\{\xi \leq r\}\,dr = \int_{-\infty}^0 1/(2-r)\,dr = +\infty.$$

Example 3.1 Assume that $\xi = (a,b)$ is an equipossible fuzzy variable. If $a \geq 0$, then for any $r \geq 0$, we have

$$\mathrm{Cr}\{\xi \geq r\} = \begin{cases} 1, & \text{if } 0 \leq r \leq a \\ 0.5, & \text{if } a < r \leq b \\ 0, & \text{if } b < r < \infty, \end{cases}$$

which implies that ξ has expected value

$$E[\xi] = \int_0^{+\infty} \mathrm{Cr}\{\xi \geq r\}\,dr = a + (b-a)/2 = (a+b)/2.$$

If $b \leq 0$, then for any $r \leq 0$, we have

$$\mathrm{Cr}\{\xi \leq r\} = \begin{cases} 1, & \text{if } b \leq r \leq 0 \\ 0.5, & \text{if } a \leq r < b \\ 0, & \text{if } -\infty < r < a, \end{cases}$$

which implies that ξ has expected value

$$E[\xi] = -\int_{-\infty}^0 \mathrm{Cr}\{\xi \leq r\}\,dr = b - (b-a)/2 = (a+b)/2.$$

Otherwise, we have $a < 0 < b$. In this case, it is easy to prove that

$$E[\xi] = \int_0^b 0.5\,dr - \int_a^0 0.5\,dr = (a+b)/2.$$

In general, an equipossible fuzzy variable (a,b) has expected value $(a+b)/2$.

Example 3.2 Assume that ξ is a simple fuzzy variable taking district values in $\{x_1, x_2, \ldots, x_m\}$. If ξ has the following credibility function

$$\nu(x) = \begin{cases} \nu_1, & \text{if } x = x_1 \\ \nu_2, & \text{if } x = x_2 \\ \cdots & \cdots \\ \nu_m, & \text{if } x = x_m, \end{cases}$$

then it has the expected value

$$E[\xi] = \sum_{i=1}^m w_i x_i \tag{3.2}$$

where for each $1 \leq i \leq m$, the weight is given by

$$w_i = \max_{x_j \leq x_i} v_j \wedge 0.5 - \max_{x_j < x_i} v_j \wedge 0.5 + \max_{x_j \geq x_i} v_j \wedge 0.5 - \max_{x_j > x_i} v_j \wedge 0.5.$$

For example, if simple fuzzy variable ξ has the following credibility function

$$v(x) = \begin{cases} 0.3, & \text{if } x = 1 \\ 0.4, & \text{if } x = 2 \\ 0.6, & \text{if } x = 3 \\ 0.2, & \text{if } x = 4, \end{cases}$$

then it is easy to prove that $w_1 = 0.3$, $w_2 = 0.1$, $w_3 = 0.4$, $w_4 = 0.2$, and

$$E[\xi] = 0.3 \times 1 + 0.1 \times 2 + 0.4 \times 3 + 0.2 \times 4 = 2.5.$$

Especially, if both $\{x_i\}$ and $\{v_i\}$ are increasing sequences, we have

$$w_1 = v_1, \qquad w_i = v_i - v_{i-1}, \quad i = 2, 3, \ldots, m-1, \qquad w_m = 1 - v_{m-1}.$$

Theorem 3.1 (Expected Value Linearity Theorem (Liu 2004)) *Let ξ and η be two independent fuzzy variables with finite expected values. Then for any real numbers a and b, we have*

$$E[a\xi + b\eta] = aE[\xi] + bE[\eta]. \tag{3.3}$$

Proof Step 1: We first prove $E[a\xi] = aE[\xi]$ for any real number a. If $a = 0$, then the equation holds trivially. If $a > 0$, we have

$$\begin{aligned} E[a\xi] &= \int_0^\infty \mathrm{Cr}\{a\xi \geq r\}\,\mathrm{d}r - \int_{-\infty}^0 \mathrm{Cr}\{a\xi \leq r\}\,\mathrm{d}r \\ &= \int_0^\infty \mathrm{Cr}\{\xi \geq r/a\}\,\mathrm{d}r - \int_{-\infty}^0 \mathrm{Cr}\{\xi \leq r/a\}\,\mathrm{d}r \\ &= a \int_0^\infty \mathrm{Cr}\{\xi \geq s\}\,\mathrm{d}s - a \int_{-\infty}^0 \mathrm{Cr}\{\xi \leq s\}\,\mathrm{d}s \\ &= aE[\xi]. \end{aligned}$$

Similarly, if $a < 0$, it follows from Definition 3.1 that

$$\begin{aligned} E[a\xi] &= \int_0^\infty \mathrm{Cr}\{a\xi \geq r\}\,\mathrm{d}r - \int_{-\infty}^0 \mathrm{Cr}\{a\xi \leq r\}\,\mathrm{d}r \\ &= \int_0^\infty \mathrm{Cr}\{\xi \leq r/a\}\,\mathrm{d}r - \int_{-\infty}^0 \mathrm{Cr}\{\xi \geq r/a\}\,\mathrm{d}r \\ &= -a \int_{-\infty}^0 \mathrm{Cr}\{\xi \leq s\}\,\mathrm{d}s + a \int_0^\infty \mathrm{Cr}\{\xi \geq s\}\,\mathrm{d}s \\ &= aE[\xi]. \end{aligned}$$

Step 2: We prove that $E[\xi + b] = E[\xi] + b$ for any real number b. If $b \geq 0$, according to the duality axiom of credibility measure, we have

$$E[\xi + b] = \int_0^\infty \mathrm{Cr}\{\xi + b \geq r\}\,\mathrm{d}r - \int_{-\infty}^0 \mathrm{Cr}\{\xi + b \leq r\}\,\mathrm{d}r$$

$$= \int_0^\infty \mathrm{Cr}\{\xi \geq r - b\}\,\mathrm{d}r - \int_{-\infty}^0 \mathrm{Cr}\{\xi \leq r - b\}\,\mathrm{d}r$$

$$= \int_{-b}^\infty \mathrm{Cr}\{\xi \geq s\}\,\mathrm{d}s - \int_{-\infty}^{-b} \mathrm{Cr}\{\xi \leq s\}\,\mathrm{d}s$$

$$= \int_{-b}^0 \mathrm{Cr}\{\xi \geq s\}\,\mathrm{d}s + \int_{-b}^0 \mathrm{Cr}\{\xi \leq s\}\,\mathrm{d}s + E[\xi]$$

$$= E[\xi] + b.$$

Similarly, in case of $b < 0$, we have

$$E[\xi + b] = E[\xi] - \int_0^{-b} \mathrm{Cr}\{\xi \geq s\}\,\mathrm{d}s - \int_0^{-b} \mathrm{Cr}\{\xi \leq s\}\,\mathrm{d}s = E[\xi] + b.$$

Step 3: We prove that $E[\xi + \eta] = E[\xi] + E[\eta]$ when both ξ and η are simple fuzzy variables with the following credibility functions,

$$\mu(x) = \begin{cases} \mu_1, & \text{if } x = a_1 \\ \mu_2, & \text{if } x = a_2 \\ \cdots & \cdots \\ \mu_m, & \text{if } x = a_m, \end{cases} \qquad \nu(x) = \begin{cases} \nu_1, & \text{if } x = b_1 \\ \nu_2, & \text{if } x = b_2 \\ \cdots & \cdots \\ \nu_n, & \text{if } x = b_n. \end{cases}$$

It is clear that $\xi + \eta$ is also a simple fuzzy variable taking values $a_i + b_j$ with credibility $\mu_i \wedge \nu_j$ for $i = 1, 2, \ldots, m$, $j = 1, 2, \ldots, n$. Now we define

$$w_i' = \max_{a_k \leq a_i} \mu_k \wedge 0.5 - \max_{a_k < a_i} \mu_k \wedge 0.5 + \max_{a_k \geq a_i} \mu_k \wedge 0.5 - \max_{a_k > a_i} \mu_k \wedge 0.5,$$

$$w_j'' = \max_{b_l \leq b_j} \nu_l \wedge 0.5 - \max_{b_l < b_j} \nu_l \wedge 0.5 + \max_{b_l \geq b_j} \nu_l \wedge 0.5 - \max_{b_l > b_j} \nu_l \wedge 0.5,$$

$$w_{ij} = \max_{a_k + b_l \leq a_i + b_j} \mu_k \wedge \nu_l \wedge 0.5 - \max_{a_k + b_l < a_i + b_j} \mu_k \wedge \nu_l \wedge 0.5$$

$$+ \max_{a_k + b_l \geq a_i + b_j} \mu_k \wedge \nu_l \wedge 0.5 - \max_{a_k + b_l > a_i + b_j} \mu_k \wedge \nu_l \wedge 0.5.$$

It is easy to prove that

$$w_i' = \sum_{j=1}^n w_{ij}, \qquad w_j'' = \sum_{i=1}^m w_{ij}.$$

for all $i = 1, 2, \ldots, m$, $j = 1, 2, \ldots, n$. If $\{a_i\}$, $\{b_j\}$ and $\{a_i + b_j\}$ are sequences consisting of distinct elements, then

$$E[\xi] = \sum_{i=1}^{m} a_i w_i', \qquad E[\eta] = \sum_{j=1}^{n} b_j w_j'', \qquad E[\xi + \eta] = \sum_{i=1}^{m} \sum_{j=1}^{n} (a_i + b_j) w_{ij}.$$

Thus $E[\xi + \eta] = E[\xi] + E[\eta]$. Otherwise, we may give them a small perturbation such that they are distinct, and then prove the linearity by letting the perturbation tend to zero.

Step 4: We prove that $E[\xi + \eta] = E[\xi] + E[\eta]$ when ξ and η are fuzzy variables such that

$$\lim_{y \uparrow 0} \mathrm{Cr}\{\xi \le y\} \le 0.5 \le \mathrm{Cr}\{\xi \le 0\}, \qquad \lim_{y \uparrow 0} \mathrm{Cr}\{\eta \le y\} \le 0.5 \le \mathrm{Cr}\{\eta \le 0\}. \qquad (3.4)$$

We define sequences of simple fuzzy variables $\{\xi_i\}$ and $\{\eta_i\}$ satisfying

$$\mathrm{Cr}\{\xi_i \le r\} \uparrow \mathrm{Cr}\{\xi \le r\}, \quad \mathrm{Cr}\{\eta_i \le r\} \uparrow \mathrm{Cr}\{\eta \le r\}, \quad \text{if } r \le 0,$$

$$\mathrm{Cr}\{\xi_i \ge r\} \uparrow \mathrm{Cr}\{\xi \ge r\}, \quad \mathrm{Cr}\{\eta_i \ge r\} \uparrow \mathrm{Cr}\{\eta \ge r\}, \quad \text{if } r \ge 0.$$

It is clear that $\{\xi_i + \eta_i\}$ is a sequence of simple fuzzy variables. Furthermore, when $r \le 0$, it follows from (3.4) that

$$\lim_{i \to \infty} \mathrm{Cr}\{\xi_i + \eta_i \le r\} = \lim_{i \to \infty} \sup_{x \le 0, y \le 0, x + y \le r} \left(\mathrm{Cr}\{\xi_i \le x\} \wedge \mathrm{Cr}\{\eta_i \le y\} \right)$$

$$= \sup_{x \le 0, y \le 0, x + y \le r} \lim_{i \to \infty} \left(\mathrm{Cr}\{\xi_i \le x\} \wedge \mathrm{Cr}\{\eta_i \le y\} \right)$$

$$= \sup_{x \le 0, y \le 0, x + y \le r} \left(\mathrm{Cr}\{\xi \le x\} \wedge \mathrm{Cr}\{\eta \le y\} \right)$$

$$= \mathrm{Cr}\{\xi + \eta \le r\}.$$

A similar way may prove that

$$\mathrm{Cr}\{\xi_i + \eta_i \ge r\} \uparrow \mathrm{Cr}\{\xi + \eta \ge r\}, \quad \text{if } r \ge 0.$$

Since the expected values $E[\xi]$ and $E[\eta]$ exist, we have

$$\lim_{i \to \infty} \int_0^\infty \mathrm{Cr}\{\xi_i \ge r\} \, dr = \int_0^\infty \mathrm{Cr}\{\xi \ge r\} \, dr$$

$$\lim_{i \to \infty} \int_{-\infty}^0 \mathrm{Cr}\{\xi_i \le r\} \, dr = \int_{-\infty}^0 \mathrm{Cr}\{\xi \le r\} \, dr$$

$$\lim_{i \to \infty} \int_0^\infty \mathrm{Cr}\{\eta_i \ge r\} \, dr = \int_0^\infty \mathrm{Cr}\{\eta \ge r\} \, dr$$

$$\lim_{i \to \infty} \int_{-\infty}^0 \mathrm{Cr}\{\eta_i \le r\} \, dr = \int_{-\infty}^0 \mathrm{Cr}\{\eta \le r\} \, dr$$

$$\lim_{i \to \infty} \int_0^\infty \text{Cr}\{\xi_i + \eta_i \geq r\}\, dr = \int_0^\infty \text{Cr}\{\xi + \eta \geq r\}\, dr$$

$$\lim_{i \to \infty} \int_{-\infty}^0 \text{Cr}\{\xi_i + \eta_i \leq r\}\, dr = \int_{-\infty}^0 \text{Cr}\{\xi + \eta \leq r\}\, dr.$$

It follows from step 3 that $E[\xi + \eta] = E[\xi] + E[\eta]$.

Step 5: We prove that $E[\xi + \eta] = E[\xi] + E[\eta]$ holds for arbitrary fuzzy variables ξ and η. Since they have finite expected values, there exist two real numbers c and d such that

$$\lim_{y \uparrow 0} \text{Cr}\{\xi + c \leq y\} \leq 0.5 \leq \text{Cr}\{\xi + c \leq 0\},$$

$$\lim_{y \uparrow 0} \text{Cr}\{\eta + d \leq y\} \leq 0.5 \leq \text{Cr}\{\eta + d \leq 0\}.$$

It follows from steps 2 and 4 that

$$\begin{aligned}
E[\xi + \eta] &= E\big[(\xi + c) + (\eta + d) - c - d\big] \\
&= E\big[(\xi + c) + (\eta + d)\big] - c - d \\
&= E[\xi + c] + E[\eta + d] - c - d \\
&= E[\xi] + E[\eta].
\end{aligned}$$

Step 6: Finally, for any $a, b \in \Re$, it follows from steps 1, 2 and 5 that

$$E[a\xi + b\eta] = E[a\xi] + E[b\eta] = a E[\xi] + b E[\eta].$$

The theorem is proved. □

Remark 3.5 The expected value linearity theorem may be not true if fuzzy variables ξ and η are not independent. For example, take a credibility space $(\Theta, \mathcal{A}, \text{Cr})$ to be $\{\theta_1, \theta_2, \theta_3\}$ with $\text{Cr}\{\theta_1\} = 0.7$, $\text{Cr}\{\theta_2\} = 0.3$ and $\text{Cr}\{\theta_3\} = 0.2$. Define two fuzzy variables

$$\xi_1(\theta) = \begin{cases} 1, & \text{if } \theta = \theta_1 \\ 0, & \text{if } \theta = \theta_2 \\ 2, & \text{if } \theta = \theta_3, \end{cases} \qquad \eta_1(\theta) = \begin{cases} 0, & \text{if } \theta = \theta_1 \\ 0, & \text{if } \theta = \theta_2 \\ 3, & \text{if } \theta = \theta_3. \end{cases}$$

It is easy to prove that $E[\xi_1] = 0.9$, $E[\eta_1] = 0.8$, and $E[\xi_1 + \eta_1] = 1.9$, which implies that

$$E[\xi_1 + \eta_1] > E[\xi_1] + E[\eta_1].$$

On the other hand, if we define

$$\xi_2(\theta) = \begin{cases} 0, & \text{if } \theta = \theta_1 \\ 1, & \text{if } \theta = \theta_2 \\ 2, & \text{if } \theta = \theta_3, \end{cases} \qquad \eta_2(\theta) = \begin{cases} 0, & \text{if } \theta = \theta_1 \\ 3, & \text{if } \theta = \theta_2 \\ 1, & \text{if } \theta = \theta_3, \end{cases}$$

we have $E[\xi_2] = 0.5$, $E[\eta_2] = 0.9$, and $E[\xi_2 + \eta_2] = 1.2$, which implies that

$$E[\xi_2 + \eta_2] < E[\xi_2] + E[\eta_2].$$

Example 3.3 For a triangular fuzzy variable $\xi = (a, b, c)$, we will prove that its expected value is

$$E[\xi] = (a + 2b + c)/4. \tag{3.5}$$

Especially, if ξ is a symmetric fuzzy variable with $b - a = c - b$, we have $E[\xi] = b$. First, we assume $a \geq 0$. For any $r \geq 0$, according to the credibility inversion theorem, we have

$$\text{Cr}\{\xi \geq r\} = \begin{cases} 1, & \text{if } 0 \leq r \leq a \\ (2b - a - r)/2(b - a), & \text{if } a < r \leq b \\ (c - r)/2(c - b), & \text{if } b < r \leq c \\ 0, & \text{if } c < r < +\infty. \end{cases}$$

Then it follows from the definition of expected value that

$$E[\xi] = \int_0^c \text{Cr}\{\xi \geq r\} \, dr = (a + 2b + c)/4.$$

In case of $a < 0$, we denote $\eta = \xi - a = (0, b - a, c - a)$. Based on above analysis, we have

$$E[\eta] = \big(0 + 2(b - a) + (c - a)\big)/4.$$

According to the expected value linearity theorem, it is easy to prove that

$$E[\xi] = E[\eta] + a = (a + 2b + c)/4.$$

Example 3.4 For a trapezoidal fuzzy variable $\xi = (a, b, c, d)$, we will prove that its expected value is

$$E[\xi] = (a + b + c + d)/4.$$

Especially, if ξ is a symmetric fuzzy variable with $b - a = d - c$, we have

$$E[\xi] = (a + d)/2 = (b + c)/2.$$

First, we assume $a \geq 0$. For any $r \geq 0$, according to the credibility inversion theorem, it is easy to prove that

$$\text{Cr}\{\xi \geq r\} = \begin{cases} 1, & \text{if } 0 \leq r \leq a \\ (2b - a - r)/2(b - a), & \text{if } a < r \leq b \\ 0.5, & \text{if } b < r \leq c \\ (d - r)/2(d - c), & \text{if } c < r \leq d \\ 0, & \text{if } d < r < +\infty. \end{cases}$$

It follows from Definition 3.1 that ξ has the expected value

$$E[\xi] = \int_0^d \mathrm{Cr}\{\xi \geq r\}\,dr = (a+b+c+d)/4.$$

In case of $a < 0$, we define $\eta = \xi - a = (0, b-a, c-a, d-a)$. Based on above analysis, we have

$$E[\eta] = \big(0 + (b-a) + (c-a) + (d-a)\big)/4.$$

According to the expected value linearity theorem, it is easy to prove that

$$E[\xi] = E[\eta] + a = (a+b+c+d)/4.$$

Example 3.5 For an exponential fuzzy variable ξ with credibility function

$$v(x) = 1/\big(1 + \exp(\pi x/\sqrt{6}m)\big), \quad \forall x \geq 0,$$

it follows from the credibility inversion theorem that

$$\mathrm{Cr}\{\xi \geq r\} = v(r)$$

for any $r \geq 0$. Then according to Definition 3.1, we have

$$E[\xi] = \int_0^\infty v(r)\,dr = \delta_1 m$$

where $\delta_1 = (\sqrt{6}\ln 2)/\pi$.

Example 3.6 For a normal fuzzy variable ξ with credibility function

$$v(x) = 1/\big(1 + \exp(\pi|x - e|/\sqrt{6}\sigma)\big), \quad \forall x \in \mathfrak{R},$$

according to the credibility inversion theorem, we have

$$\mathrm{Cr}\{\xi \geq r\} = v(r), \quad \forall r \geq e,$$
$$\mathrm{Cr}\{\xi \leq r\} = v(r), \quad \forall r \leq e.$$

Then it follows from Definition 3.1 that

$$E[\xi] = \int_0^e \big(\mathrm{Cr}\{\xi \geq r\} + \mathrm{Cr}\{\xi \leq r\}\big)\,dr + \int_e^\infty v(r)\,dr - \int_{-\infty}^e v(r)\,dr$$
$$= \int_0^e \big(\mathrm{Cr}\{\xi \geq r\} + \mathrm{Cr}\{\xi \leq r\}\big)\,dr$$
$$= e.$$

In fact, for any continuous fuzzy variable with a symmetric and unimodal credibility function, a similar way may prove that its expected value is the symmetric center.

Variance

The variance provides a spread degree of the fuzzy variable around its expected value. A small value of variance indicates that the fuzzy variable is tightly concentrated around its expected value; and a large value of variance indicates that the fuzzy variable has a wide spread around its expected value.

Definition 3.2 (Liu 2004) Let ξ be a fuzzy variable with finite expected value. Then its variance is defined as

$$V[\xi] = E\big[(\xi - E[\xi])^2\big]. \tag{3.6}$$

This definition tells us that the variance is just the expected value of the nonnegative fuzzy variable $(\xi - E[\xi])^2$, that is,

$$V[\xi] = \int_0^\infty \mathrm{Cr}\big\{(\xi - E[\xi])^2 \geq r\big\}\,\mathrm{d}r. \tag{3.7}$$

Example 3.7 Let $\xi = (a, b)$ be an equipossible fuzzy variable. Then we have $E[\xi] = (a + b)/2$. For any $r \geq 0$, according to the credibility inversion theorem, we have

$$\mathrm{Cr}\big\{(\xi - E[\xi])^2 \geq r\big\} = \begin{cases} 0.5, & \text{if } r \leq (b - a)^2/4 \\ 0, & \text{if } r > (b - a)^2/4. \end{cases}$$

Thus, it follows from (3.7) that its variance is

$$V[\xi] = (b - a)^2/8.$$

Example 3.8 Let $\xi = (a, b, c)$ be a triangular fuzzy variable. Define $\alpha = \max\{b - a, c - b\}$ and $\beta = \min\{b - a, c - b\}$. It is easy to prove that

$$V[\xi] = \frac{33\alpha^3 + 21\alpha^2\beta + 11\alpha\beta^2 - \beta^3}{384\alpha}. \tag{3.8}$$

Especially, if ξ is a symmetric fuzzy variable with $\alpha = \beta$, then we have

$$V[\xi] = (c - a)^2/24.$$

Example 3.9 Let $\xi = (a, b, c, d)$ be a symmetric trapezoidal fuzzy variable. It follows from (3.7) that its variance is

$$V[\xi] = \big((d - a)^2 + (d - a)(c - b) + (c - b)^2\big)/24.$$

Example 3.10 Let $\xi = EXP(m)$ be an exponential fuzzy variable. Example 3.5 has shown that it has an expected value

$$e = (\sqrt{6}m \ln 2)/\pi.$$

Then it follows from (3.7) that

$$V[\xi] = \int_0^\infty \mathrm{Cr}\big\{(\xi - e)^2 \geq r\big\}\,\mathrm{d}r$$

$$= \int_0^{e^2} \mathrm{Cr}\{\xi \leq e - \sqrt{r}\}\,\mathrm{d}r + \int_{e^2}^\infty \mathrm{Cr}\{\xi \geq e + \sqrt{r}\}\,\mathrm{d}r$$

$$= \int_0^e \mathrm{Cr}\{\xi \leq e - s\}\,\mathrm{d}s^2 + \int_e^\infty \mathrm{Cr}\{\xi \geq e + s\}\,\mathrm{d}s^2$$

$$= \int_0^e \frac{4s}{2 + \exp(\pi s/\sqrt{6}m)}\,\mathrm{d}s + \int_e^\infty \frac{2s}{1 + 2\exp(\pi s/\sqrt{6}m)}\,\mathrm{d}s$$

$$= \delta_2 m^2$$

where δ_2 is a constant value

$$\frac{12}{\pi}\left(\int_0^{\ln 2} \frac{2s}{2 + \exp(s)}\,\mathrm{d}s + \int_{\ln 2}^\infty \frac{s}{1 + 2\exp(s)}\,\mathrm{d}s\right) \approx 2.0031.$$

Example 3.11 For a normal fuzzy variable $\xi = N(e, \sigma)$, we will prove that it has a variance σ^2. First, according to the maximality axiom, we have

$$\mathrm{Cr}\big\{(\xi - e)^2 \geq x\big\} = \mathrm{Cr}\{\xi - e \geq \sqrt{x}\}$$

for any $x \geq 0$. Then it follows from (3.7) that

$$V[\xi] = \int_0^\infty \mathrm{Cr}\{\xi - e \geq \sqrt{x}\}\,\mathrm{d}x$$

$$= \int_0^\infty 2x\mathrm{Cr}\{\xi - e \geq x\}\,\mathrm{d}x$$

$$= \int_e^\infty 2(x - e)\nu(x)\,\mathrm{d}x$$

$$= \sigma^2.$$

This example tells us that the parameter σ appearing in the normal credibility function denotes the standard variance.

Theorem 3.2 (Liu 2004) *Let ξ be a fuzzy variable with finite expected value. For any real numbers a and b, we have*

$$V[a\xi + b] = a^2 V[\xi]. \tag{3.9}$$

Proof Suppose that ξ has finite expected value e. According to the expected value linearity theorem, we have $E[a\xi + b] = ae + b$. Then it follows from the definition of variance that

$$V[a\xi + b] = E\big[(a\xi + b - (ae + b))^2\big] = a^2 E\big[(\xi - e)^2\big] = a^2 V[\xi].$$

The proof is complete.   □

Theorem 3.3 (Liu 2004) *Suppose that ξ is a fuzzy variable with finite expected value e. Then we have $V[\xi] = 0$ if and only if $\mathrm{Cr}\{\xi = e\} = 1$, i.e., ξ is essentially the constant number e.*

Proof We first assume $V[\xi] = 0$. It follows from (3.7) that

$$\int_0^\infty \mathrm{Cr}\big\{(\xi - e)^2 \ge r\big\} \, dr = 0$$

which implies $\mathrm{Cr}\{(\xi - e)^2 \ge r\} = 0$ for any $r > 0$. Hence we have

$$\mathrm{Cr}\big\{(\xi - e)^2 = 0\big\} = 1.$$

That is, $\mathrm{Cr}\{\xi = e\} = 1$. Conversely, we assume $\mathrm{Cr}\{\xi = e\} = 1$. Then we immediately have $\mathrm{Cr}\{(\xi - e)^2 \ge r\} = 0$ for any $r > 0$. Thus

$$V[\xi] = \int_0^\infty \mathrm{Cr}\big\{(\xi - e)^2 \ge r\big\} \, dr = 0.$$

The theorem is proved.   □

Let ξ be a fuzzy variable that takes values in $[a, b]$, but whose credibility function is otherwise arbitrary. If its expected value is given, above theorem tells us that its minimum variance is zero. On the other hand, what is the possible maximum variance? The following maximum variance theorem will answer this question.

Theorem 3.4 (Li et al. 2010c) *Let f be a convex function on $[a, b]$, and let ξ be a fuzzy variable taking values in $[a, b]$. Then we have*

$$E\big[f(\xi)\big] \le \frac{b - E[\xi]}{b - a} f(a) + \frac{E[\xi] - a}{b - a} f(b). \tag{3.10}$$

Proof Suppose that ξ is a fuzzy variable defined on the credibility space $(\Theta, \mathcal{A}, \mathrm{Cr})$. For each $\theta \in \Theta$, we have $a \le \xi(\theta) \le b$ and

$$\xi(\theta) = \frac{b - \xi(\theta)}{b - a} a + \frac{\xi(\theta) - a}{b - a} b.$$

It follows from the convexity of f that

$$f\big(\xi(\theta)\big) \le \frac{b - \xi(\theta)}{b - a} f(a) + \frac{\xi(\theta) - a}{b - a} f(b).$$

Taking expected values on both sides, we obtain the inequality.   □

Theorem 3.5 (Maximum Variance Theorem (Li et al. 2010c)) *Let ξ be a fuzzy variable that takes values in $[a, b]$ and has finite expected value e. Then we have*

$$V[\xi] \le (e - a)(b - e) \tag{3.11}$$

and the equality holds if the fuzzy variable ξ has credibility function

$$v(x) = \begin{cases} (b - e)/(b - a), & \text{if } x = a \\ (e - a)/(b - a), & \text{if } x = b. \end{cases} \tag{3.12}$$

Proof First, inequality (3.11) follows from Theorem 3.4 immediately by defining $f(x) = (x - e)^2$. In addition, it is also easy to verify that the fuzzy variable determined by credibility function (3.12) has variance $(e - a)(b - e)$. The theorem is proved. □

Skewness

Skewness is used to measure the preference that fuzzy variable ξ takes a larger value than its expected value.

Definition 3.3 (Li et al. 2010a) Let ξ be a fuzzy variable with finite expected value. Then its skewness is defined as

$$S[\xi] = E\left[(\xi - E[\xi])^3\right]. \tag{3.13}$$

Example 3.12 Let $\xi = (a, b)$ be an equipossible fuzzy variable with $a = 0$. Then it has expected value $E[\xi] = b/2$ and skewness

$$S[\xi] = \int_0^\infty \mathrm{Cr}\left\{(\xi - b/2)^3 \ge r\right\} dr - \int_{-\infty}^0 \mathrm{Cr}\left\{(\xi - b/2)^3 \le r\right\} dr$$

$$= \int_0^\infty 3r^2 \mathrm{Cr}\{\xi - b/2 \ge r\} dr - \int_{-\infty}^0 3r^2 \mathrm{Cr}\{\xi - b/2 \le r\} dr$$

$$= \int_0^{b/2} 1.5r^2 dr - \int_{-b/2}^0 1.5r^2 dr$$

$$= 0.$$

If $a \ne 0$, it follows from the expected value linearity theorem that

$$S[\xi] = E\left[(\xi - E[\xi])^3\right] = E\left[((\xi - a) - E[\xi - a])^3\right] = 0.$$

Example 3.13 Let $\xi = (a, b, c)$ be a triangular fuzzy variable. In what follows, we will prove that

$$S[\xi] = (c - a)^2\left((c - b) - (b - a)\right)/32. \tag{3.14}$$

Denote $e = (a + 2b + c)/4$. For simplicity, we assume that $c - b \geq b - a$ and $a > 0$. In this case, it follows from the definition of skewness that

$$
\begin{aligned}
S[\xi] &= \int_0^{c-e} \mathrm{Cr}\{\xi - e \geq r\} \, dr^3 - \int_{a-e}^0 \mathrm{Cr}\{\xi - e \leq r\} \, dr^3 \\
&= \int_0^{c-e} \frac{c - e - r}{2(c - b)} \, dr^3 - \int_{a-e}^{b-e} \frac{e + r - a}{2(b - a)} \, dr^3 - \int_{b-e}^0 \frac{c - 2b + e + r}{2(c - b)} \, dr^3 \\
&= \frac{(c - e)^4 - (b - e)^4}{8(c - b)} + \frac{(b - e)^4 - (a - e)^4}{8(b - a)} \\
&= (c - a)^2\big((c - b) - (b - a)\big)/32.
\end{aligned}
$$

It is clear that the skewness takes value zero if ξ is a symmetric triangular fuzzy variable with $c - b = b - a$.

Example 3.14 A trapezoidal fuzzy variable $\xi = (a, b, c, d)$ has a skewness

$$
S[\xi] = (d + c - a - b)\big((d - c)^2 - (b - a)^2\big)/32. \tag{3.15}
$$

For simplicity, we only consider the case of $a > 0$ and $b \leq e \leq c$ where $e = (a + b + c + d)/4$ is the expected value of ξ. It follows from (3.13) that

$$
\begin{aligned}
S[\xi] &= \int_0^{d-e} \mathrm{Cr}\{\xi - e \geq r\} \, dr^3 - \int_{a-e}^0 \mathrm{Cr}\{\xi - e \leq r\} \, dr^3 \\
&= \int_{c-e}^{d-e} \frac{d - e - r}{2(d - c)} \, dr^3 - \int_{a-e}^{b-e} \frac{e + r - a}{2(b - a)} \, dr^3 + \frac{(c - e)^3}{2} + \frac{(b - e)^3}{2} \\
&= \frac{(d - e)^4 - (c - e)^4}{8(d - c)} + \frac{(b - e)^4 - (a - e)^4}{8(b - a)} \\
&= (d + c - a - b)\big((d - c)^2 - (b - a)^2\big)/32.
\end{aligned}
$$

It is clear that the skewness takes value zero if ξ is a symmetric trapezoidal fuzzy variable with $d - c = b - a$.

Example 3.15 Let $\xi = N(e, \sigma)$ be a normal fuzzy variable. For any real number r, it follows from the credibility inversion theorem that

$$
\mathrm{Cr}\{\xi \leq r\} = \frac{1}{1 + \exp(\pi(e - r)/\sqrt{6}\sigma)},
$$

$$
\mathrm{Cr}\{\xi \geq r\} = \frac{1}{1 + \exp(\pi(r - e)/\sqrt{6}\sigma)}.
$$

Then it follows form (3.13) that

$$S[\xi] = \int_0^{+\infty} \mathrm{Cr}\{\xi - e \geq r\}\, dr^3 - \int_{-\infty}^0 \mathrm{Cr}\{\xi - e \leq r\}\, dr^3$$

$$= \int_0^{+\infty} \frac{1}{1 + \exp(\pi r/\sqrt{6}\sigma)}\, dr^3 - \int_{-\infty}^0 \frac{1}{1 + \exp(-\pi r/\sqrt{6}\sigma)}\, dr^3$$

$$= \int_0^{+\infty} \frac{1}{1 + \exp(\pi r/\sqrt{6}\sigma)}\, dr^3 - \int_0^{+\infty} \frac{1}{1 + \exp(\pi r/\sqrt{6}\sigma)}\, dr^3$$

$$= 0.$$

Example 3.16 Let $\xi = EXP(m)$ be an exponential fuzzy variable. For any real number r, it follows from the credibility inversion theorem that

$$\mathrm{Cr}\{\xi \leq r\} = 1 - \frac{1}{1 + \exp(\pi r/\sqrt{6}m)}, \qquad \mathrm{Cr}\{\xi \geq r\} = \frac{1}{1 + \exp(\pi r/\sqrt{6}m)}.$$

Then it follows form (3.13) that

$$S[\xi] = \frac{18\sqrt{6}}{\pi^3}\left(\int_{\ln 2}^{+\infty} \frac{(r - \ln 2)^2}{1 + \exp(r)}\, dr - \int_0^{\ln 2} \frac{(r - \ln 2)^2}{1 + \exp(-r)}\, dr\right)m^3 = \delta_3 m^3.$$

It is calculated that $\delta_3 \approx 2.914$, which tells us that the exponential fuzzy variable always has a positive skewness.

Theorem 3.6 (Li et al. 2010a) *Let ξ be a fuzzy variable with finite expected value. For any real numbers a and b, we have*

$$S[a\xi + b] = a^3 S[\xi].\tag{3.16}$$

Proof First, it follows from the expected value linearity theorem that $E[a\xi + b] = aE[\xi] + b$. Then according to Definition 3.3, we have

$$S[a\xi + b] = E\left[\left((a\xi + b) - (aE[\xi] + b)\right)^3\right]$$

$$= a^3 E\left[(\xi - E[\xi])^3\right]$$

$$= a^3 S[\xi].$$

The proof is complete.   □

Theorem 3.7 (Li et al. 2010a) *Let ξ be a symmetric fuzzy variable with finite expected value. Then we have $S[\xi] = 0$.*

Proof Suppose that ξ has a credibility function ν. Since ξ is a symmetric fuzzy variable, there is a real number e such that $\nu(x - e) = \nu(x + e)$ for any $x \in \mathfrak{R}$.

According to the credibility inversion theorem, we have

$$\mathrm{Cr}\{\xi \geq e + r\} = \mathrm{Cr}\{\xi \leq e - r\} \tag{3.17}$$

for any $r \in \Re$. We will prove that $E[\xi] = e$. In fact, it follows from the definition of expected value that

$$
\begin{aligned}
E[\xi] &= \int_0^{+\infty} \mathrm{Cr}\{\xi \geq r\}\,dr - \int_{-\infty}^0 \mathrm{Cr}\{\xi \leq r\}\,dr \\
&= \int_{-e}^{+\infty} \mathrm{Cr}\{\xi \geq r + e\}\,dr - \int_{-\infty}^e \mathrm{Cr}\{\xi \leq r - e\}\,dr \\
&= \int_{-e}^{+\infty} \mathrm{Cr}\{\xi \geq r + e\}\,dr - \int_e^{+\infty} \mathrm{Cr}\{\xi \leq e - r\}\,dr \\
&= \int_{-e}^0 \mathrm{Cr}\{\xi \geq r + e\}\,dr + \int_0^e \mathrm{Cr}\{\xi \leq e - r\}\,dr \\
&= \int_0^e \left(\mathrm{Cr}\{\xi \geq e - r\} + \mathrm{Cr}\{\xi < e - r\} \right) dr.
\end{aligned}
$$

Then it follows from the duality axiom that $E[\xi] = e$. Finally, it follows from the definition of skewness and (3.17) that

$$
\begin{aligned}
S[\xi] &= \int_0^{+\infty} \mathrm{Cr}\{\xi \geq r + e\}\,dr^3 - \int_{-\infty}^0 \mathrm{Cr}\{\xi \leq r + e\}\,dr^3 \\
&= \int_0^{+\infty} \mathrm{Cr}\{\xi \geq r + e\}\,dr^3 - \int_0^{+\infty} \mathrm{Cr}\{\xi \leq e - r\}\,dr^3 \\
&= 0.
\end{aligned}
$$

The proof is complete.   □

Moment

Definition 3.4 (Liu 2007) Let ξ be a fuzzy variable with finite expected value e, and let k be a positive integer. Then

(a) the expected value $E[\xi^k]$ is called the kth moment;
(b) the expected value $E[|\xi|^k]$ is called the kth absolute moment;
(c) the expected value $E[(\xi - e)^k]$ is called the kth central moment;
(d) the expected value $E[|\xi - e|^k]$ is called the kth absolute central moment.

Note that the first moment is the expected value, the second central moment is the variance, and the third central moment is the skewness.

Example 3.17 Let $\xi = EXP(m)$ be an exponential fuzzy variable. It follows from the credibility inversion theorem that

$$\mathrm{Cr}\{\xi \geq r\} = \nu(r)$$

for all $r > 0$. Then according to Definition 3.1, we have

$$E[\xi^2] = \int_0^\infty 2r\,\mathrm{Cr}\{\xi \geq r\}\,dr = \int_0^\infty \frac{2r}{1 + \exp(\pi r/(\sqrt{6}m))}\,dr = m^2.$$

This example tells us that the parameter m appearing in the exponential credibility function denotes the standard second moment.

3.2 Expected Value Model

As one of the most important credibilistic mappings, expected value is used to denote the average value for each fuzzy quantity. If the decision-maker would like to obtain a maximum average objective value subject to a set of expected constraints, we have the following credibilistic programming model (Liu and Liu 2002),

$$\begin{cases} \max & E[f(x, \xi)] \\ \text{s.t.} & E[g_i(x, \xi)] \leq 0, \quad i = 1, 2, \ldots, n. \end{cases} \tag{3.18}$$

For simplicity, we call it an *expected value model*.

Remark 3.6 The concepts of feasible solution, local optimal solution, and global optimal solution are given by Definitions 2.6, 2.7, and 2.8.

Theorem 3.8 *Assume that $\xi_1, \xi_2, \ldots, \xi_m$ are independent fuzzy variables. If the objective function and constraint functions satisfy*

$$f(x, \xi) = f_0(x) + f_1(x)\xi_1 + f_2(x)\xi_2 + \cdots + f_m(x)\xi_m,$$

$$g_i(x, \xi) = g_{i0}(x) + g_{i1}(x)\xi_1 + g_{i2}(x)\xi_2 + \cdots + g_{im}(x)\xi_m,$$

for all $i = 1, 2, \ldots, n$, then model (3.18) has the following crisp equivalent,

$$\begin{cases} \max & f(x, E[\xi]) \\ \text{s.t.} & g_i(x, E[\xi]) \leq 0, \quad i = 1, 2, \ldots, n \end{cases} \tag{3.19}$$

where $E[\xi] = (E[\xi_1], E[\xi_2], \ldots, E[\xi_m])$.

Proof It follows immediately from the expected value linearity theorem. □

In many cases, there are multiple objectives. Then we have the following multi-objective expected value model

$$
\begin{cases}
\max & \left[E\big[f_1(x,\xi)\big],\, E\big[f_2(x,\xi)\big],\, \ldots,\, E\big[f_p(x,\xi)\big]\right] \\
\text{s.t.} & E\big[g_i(x,\xi)\big] \le 0, \quad i = 1, 2, \ldots, n.
\end{cases}
\tag{3.20}
$$

Theorem 3.9 *Assume that* $\xi_1, \xi_2, \ldots, \xi_m$ *are independent fuzzy variables. If the objective functions and constraint functions satisfy*

$$
f_j(x,\xi) = f_{j0}(x) + f_{j1}(x)\xi_1 + f_{j2}(x)\xi_2 + \cdots + f_{jm}(x)\xi_m, \quad j = 1, 2, \ldots, p,
$$

$$
g_i(x,\xi) = g_{i0}(x) + g_{i1}(x)\xi_1 + g_{i2}(x)\xi_2 + \cdots + g_{im}(x)\xi_m, \quad i = 1, 2, \ldots, n,
$$

then the multi-objective expected value model has the following crisp equivalent,

$$
\begin{cases}
\max & \left[f_1\big(x, E[\xi]\big),\, f_2\big(x, E[\xi]\big),\, \ldots,\, f_p\big(x, E[\xi]\big)\right] \\
\text{s.t.} & g_i\big(x, E[\xi]\big) \le 0, \quad i = 1, 2, \ldots, n.
\end{cases}
\tag{3.21}
$$

Proof It follows immediately from the expected value linearity theorem.   □

3.3 Fuzzy Simulation

In order to solve the general fuzzy expected value models, this section introduces a fuzzy simulation technique (Liu and Liu 2002; Liu 2006) to approximate the credibilistic mapping

$$
U : x \to E\big[f(x,\xi)\big]
\tag{3.22}
$$

where f is a real valued function.

Assume that the fuzzy vector ξ has a joint credibility function ν. We first introduce the simulation method on the credibility values of fuzzy events. Generate vectors $y_1, y_2, \ldots, y_N$ randomly, and calculate the credibilities

$$
\nu_k = \nu(y_k), \quad k = 1, 2, \ldots, N.
$$

For any real number r, according to the credibility inversion theorem, the credibility $\mathrm{Cr}\{f(x,\xi) \ge r\}$ can be estimated by

$$
\begin{cases}
\max\{\nu_k \mid f(x, y_k) \ge r\}, & \text{if } \max\{\nu_k \mid f(x, y_k) \ge r\} < 0.5 \\
1 - \max\{\nu_k \mid f(x, y_k) < r\}, & \text{if } \max\{\nu_k \mid f(x, y_k) \ge r\} \ge 0.5,
\end{cases}
\tag{3.23}
$$

and the credibility $\mathrm{Cr}\{f(x,\xi) \le r\}$ can be estimated by

$$
\begin{cases}
\max\{\nu_k \mid f(x, y_k) \le r\}, & \text{if } \max\{\nu_k \mid f(x, y_k) \le r\} < 0.5 \\
1 - \max\{\nu_k \mid f(x, y_k) > r\}, & \text{if } \max\{\nu_k \mid f(x, y_k) \le r\} \ge 0.5.
\end{cases}
\tag{3.24}
$$

Then the expected value may be estimated by the following procedure.

Algorithm 3.1 (Fuzzy simulation for expected value)

Step 1. Set $e = 0$.

Step 2. Randomly generate vectors $y_1, y_2, \ldots, y_N$ and calculate the credibilities $\nu_1, \nu_2, \ldots, \nu_N$.

Step 3. Set two numbers $a = f(x, y_1) \wedge f(x, y_2) \wedge \cdots \wedge f(x, y_N)$ and $b = f(x, y_1) \vee f(x, y_2) \vee \cdots \vee f(x, y_N)$.

Step 4. Randomly generate a real number r from $[a, b]$.

Step 5. If $r \geq 0$, set $e \rightarrow e + \text{Cr}\{f(x, \xi) \geq r\}$.

Step 6. If $r \leq 0$, set $e \rightarrow e - \text{Cr}\{f(x, \xi) \leq r\}$.

Step 7. Repeat the fourth to sixth steps for N times.

Step 8. Return $E = a \vee 0 + b \wedge 0 + e \cdot (b - a)/N$.

Example 3.18 The parameter N has a great influence on the simulation accuracy. Generally speaking, a larger value can obtain a better approximation with a longer computation time. Since the accuracy and the computation time are both crucial when we apply the fuzzy simulation method to solve the expected value model, it is meaningful to find the smallest value of N which can obtain a satisfactory approximation. For the triangular fuzzy variable $\xi = (1, 2, 3)$ with expected value $E[\xi] = 1$, we perform Algorithm 3.1 by changing N from 100 to 5000 with step of 100. The simulated results are illustrated by Fig. 3.1. It is shown that when N is larger than 3000, the simulated results are stable and satisfactory.

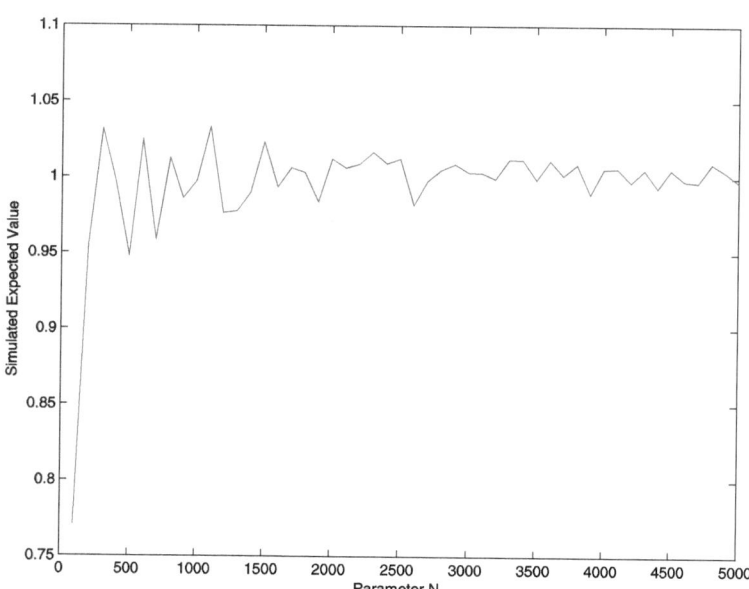

Fig. 3.1 Expected value simulation with variable parameter N

Table 3.1 Expected value simulation of different fuzzy variables

Fuzzy variables	Simulated value	Exact value	Relative error
$(0.0, 1.0)$	0.5000	0.5000	0.0000
$(-1.0, 2.0)$	0.4949	0.5000	0.0101
$(3.5, 7.8)$	5.6501	5.6500	0.0000
$(-50, -10)$	−29.9980	−30.0000	0.0000
$(-0.3, 1.8, 2.3)$	1.3998	1.4000	0.0002
$(1.5, 3.0, 4.1)$	2.8887	2.9000	0.0039
$(10, 15, 20)$	15.0652	15.0000	0.0043
$(25, 40, 50)$	38.4305	38.7500	0.0082
$(1.0, 2.0, 3.0, 4.0)$	2.4766	2.5000	0.0094
$(3.1, 4.2, 4.5, 6.0)$	4.5277	4.4500	0.0172
$(2.4, 3.6, 3.7, 5.6)$	3.6832	3.8250	0.0371
$(10, 25, 30, 45)$	27.9256	27.5000	0.0152
$N(1.5, 1.0)$	1.4857	1.5000	0.0095
$N(2.5, 1.0)$	2.3787	2.5000	0.0485
$N(3.7, 2.1)$	3.6231	3.7000	0.0208
$N(6.0, 1.3)$	6.0611	6.0000	0.0101
$E(1.3)$	0.7100	0.7026	0.0105
$E(1.0)$	0.5675	0.5404	0.0477
$E(3.5)$	1.9099	1.8916	0.0096
$E(7.6)$	4.0873	4.1074	0.0049

Example 3.19 Taking $N = 3000$, we perform Algorithm 3.1 on twenty fuzzy variables, including equipossible fuzzy variables, triangular fuzzy variables, trapezoidal fuzzy variables, exponential fuzzy variables, and normal fuzzy variables. We record the simulated values by Table 3.1, and make comparisons with the exact values. If we use s to denote the simulated value and use t to denote the exact value, the relative error is defined as

$$\delta = \left(|s - t| / \max \left(|s|, |t| \right) \right) \times 100 \,\%. \tag{3.25}$$

In Table 3.1, the last column records the relative error which ranges from 0.00 % to 4.85 % and the average value is 1.34 %. These results imply that the simulation algorithm can obtain a satisfactory approximation.

Example 3.20 Note that Algorithm 3.1 is essentially a combination of the Monte Carlo simulation and the numerical integration. Therefore, we may get different values if we perform the algorithm more than one times on the same fuzzy variable. In this example, we take $N = 3000$, and perform Algorithm 3.1 fifty times on triangular fuzzy variable $\xi = (-0.3, 1.8, 2.3)$. The simulated results are shown by Table 3.2, where the second and the fifth columns records the simulated values, and

Table 3.2 Expected value simulation on $\xi = (-0.3, 1.8, 2.3)$

No.	Simulated value	Error	No.	Simulated value	Error
1	1.2976	0.0731	26	1.4186	0.0133
2	1.4144	0.0103	27	1.5179	0.0842
3	1.4621	0.0444	28	1.3921	0.0056
4	1.3894	0.0076	29	1.3623	0.0269
5	1.4122	0.0087	30	1.5074	0.0767
6	1.4500	0.0357	31	1.3677	0.0231
7	1.4193	0.0138	32	1.4788	0.0563
8	1.3920	0.0057	33	1.3648	0.0251
9	1.3846	0.0110	34	1.3314	0.0490
10	1.4116	0.0083	35	1.3333	0.0476
11	1.3441	0.0399	36	1.4399	0.0285
12	1.3807	0.0138	37	1.3128	0.0623
13	1.3998	0.0000	38	1.4003	0.0000
14	1.3715	0.0204	39	1.3184	0.0583
15	1.3664	0.0240	40	1.3755	0.0175
16	1.4683	0.0488	41	1.3779	0.0158
17	1.4703	0.0502	42	1.3971	0.0021
18	1.4366	0.0261	43	1.4059	0.0042
19	1.4597	0.0426	44	1.4593	0.0424
20	1.3630	0.0264	45	1.3418	0.0416
21	1.3778	0.0159	46	1.4101	0.0072
22	1.3352	0.0463	47	1.3657	0.0245
23	1.4250	0.0179	48	1.3755	0.0175
24	1.4842	0.0601	49	1.2555	0.1032
25	1.3691	0.0221	50	1.4754	0.0539

the third and the sixth columns record the relative errors compared with the exact value $(-0.3 + 2 \times 1.8 + 2.3)/4 = 1.4$.

3.4 Applications

As an application, this section applies the expected value model to study the fuzzy portfolio selection problems. See Examples 2.4 and 2.5. Generally speaking, most of the investors prefer the portfolio with larger investment return and lower investment risk. Following Markowitz's work (Markowitz 1952), we can quantify the investment return by the expected value

$$E\big[f(\boldsymbol{x}, \boldsymbol{\xi})\big] = E[\xi_1 x_1 + \xi_2 x_2 + \cdots + \xi_m x_m],$$

and quantify the investment risk by the variance

$$V[f(x, \xi)] = V[\xi_1 x_1 + \xi_2 x_2 + \cdots + \xi_m x_m].$$

Furthermore, Samuelson (1970) showed that almost all investors prefer the portfolio with a larger skewness, which means that it is more possible to obtain a larger return than the expected value. Based on above analysis, we get the following multi-objective credibilistic programming model,

$$\begin{cases} \max & [E[f(x, \xi)], -V[f(x, \xi)], S[f(x, \xi)]] \\ \text{s.t.} & x_1 + x_2 + \cdots + x_m = 1 \\ & x_i \geq 0, \quad i = 1, 2, \ldots, m. \end{cases} \quad (3.26)$$

If the investor is capable of providing the preference parameters on the investment return, investment risk and skewness, then we can get the following linearly weighted compromise model

$$\begin{cases} \max & \lambda_1 E[f(x, \xi)] - \lambda_2 V[f(x, \xi)] + \lambda_3 S[f(x, \xi)] \\ \text{s.t.} & x_1 + x_2 + \cdots + x_m = 1 \\ & x_i \geq 0, \quad i = 1, 2, \ldots, m \end{cases}$$

where λ_1, λ_2, λ_3 are nonnegative real numbers. Generally speaking, a rational investor will assign the largest preference value to the most important objective, and set $\lambda_i = 0$ if he/she does not concern the objective. For example, if he/she only concern the investment return but does not care the investment risk, we may set $\lambda_1 = 0.8$, $\lambda_2 = 0$, and $\lambda_3 = 0.2$.

If the investor would like to minimize the distance between the objective values and the ideal values, i.e., the optimal values of each objective function without considering other objectives under the portfolio constraints, we can get the following ideal point compromise model

$$\begin{cases} \min & \sqrt{(E[f(x, \xi)] - e)^2 + (V[f(x, \xi)] - v)^2 + (S[f(x, \xi)] - s)^2} \\ \text{s.t.} & x_1 + x_2 + \cdots + x_m = 1 \\ & x_i \geq 0, \quad i = 1, 2, \ldots, m \end{cases}$$

where e, v and s are the ideal values of the investment return, investment risk and skewness, respectively.

When the minimal skewness level γ and the maximal risk level β are given, we have the following mean-variance-skewness model (Li et al. 2010a)

$$\begin{cases} \max & E[\xi_1 x_1 + \xi_2 x_2 + \ldots + \xi_m x_m] \\ \text{s.t.} & V[\xi_1 x_1 + \xi_2 x_2 + \cdots + \xi_m x_m] \leq \beta \\ & S[\xi_1 x_1 + \xi_2 x_2 + \cdots + \xi_m x_m] \geq \gamma \\ & x_1 + x_2 + \cdots + x_m = 1 \\ & x_i \geq 0, \quad i = 1, 2, \ldots, m \end{cases} \quad (3.27)$$

which maximizes the expected return under the risk and skewness constraints. The first variation of the mean-variance-skewness model is formulated to minimize risk under the expected return and skewness constraints

$$
\begin{cases}
\min & V[\xi_1 x_1 + \xi_2 x_2 + \cdots + \xi_m x_m] \\
\text{s.t.} & E[\xi_1 x_1 + \xi_2 x_2 + \cdots + \xi_m x_m] \geq \alpha \\
& S[\xi_1 x_1 + \xi_2 x_2 + \cdots + \xi_m x_m] \geq \gamma \\
& x_1 + x_2 + \cdots + x_m = 1 \\
& x_i \geq 0, \quad i = 1, 2, \ldots, m
\end{cases}
\tag{3.28}
$$

where α is the minimal return level. The second variation of the mean-variance-skewness model is the following,

$$
\begin{cases}
\max & S[\xi_1 x_1 + \xi_2 x_2 + \cdots + \xi_m x_m] \\
\text{s.t.} & E[\xi_1 x_1 + \xi_2 x_2 + \cdots + \xi_m x_m] \geq \alpha \\
& V[\xi_1 x_1 + \xi_2 x_2 + \cdots + \xi_m x_m] \leq \beta \\
& x_1 + x_2 + \cdots + x_m = 1 \\
& x_i \geq 0, \quad i = 1, 2, \ldots, m.
\end{cases}
\tag{3.29}
$$

In what follows, we only consider the mathematical properties and numerical examples on model (3.27). The others may be discussed similarly.

Remark 3.7 Suppose that the stock returns $\xi_i = N(e_i, \sigma_i)$, $i = 1, 2, \ldots, m$ are mutually independent normal fuzzy variables. Then it is easy to prove that the total return is also a normal fuzzy variable, which has expected value

$$
E[f(x, \xi)] = e_1 x_1 + e_2 x_2 + \cdots + e_m x_m,
$$

and has variance

$$
V[f(x, \xi)] = (\sigma_1 x_1 + \sigma_2 x_2 + \cdots + \sigma_m x_m)^2.
$$

In addition, it follows from the symmetry that the total return has skewness zero. Thus model (3.27) has the following crisp equivalent,

$$
\begin{cases}
\max & e_1 x_1 + e_2 x_2 + \cdots + e_m x_m \\
\text{s.t.} & \sigma_1 x_1 + \sigma_2 x_2 + \cdots + \sigma_m x_m \leq \sqrt{\beta} \\
& x_1 + x_2 + \cdots + x_m = 1 \\
& x_i \geq 0, \quad i = 1, 2, \ldots, m.
\end{cases}
\tag{3.30}
$$

If the stock returns $\xi_i = (a_i, b_i, c_i)$, $i = 1, 2, \ldots, m$ are mutually independent triangular fuzzy variables, then for each portfolio $(x_1, x_2, \ldots, x_m)$, the total return is a triangular fuzzy variable denoted as

$$
f(x, \xi) = \left(\sum_{i=1}^{m} a_i x_i, \sum_{i=1}^{m} b_i x_i, \sum_{i=1}^{m} c_i x_i \right).
$$

It follows from (3.5), (3.8) and (3.14) that the mean-variance-skewness model (3.27) has the following crisp equivalent,

$$
\begin{cases}
\max \ \displaystyle\sum_{i=1}^{m} x_i(a_i + 2b_i + c_i) \\[2ex]
\text{s.t.} \ \left| \displaystyle\sum_{i=1}^{m}\sum_{j=1}^{m}\sum_{k=1}^{m} 6\big((c_i - b_i)(c_j - b_j) + (b_i - a_i)(b_j - a_j)\big)(c_k + a_k - 2b_k)x_i x_j x_k \right. \\[2ex]
\qquad + \displaystyle\sum_{i=1}^{m}\sum_{j=1}^{m}\sum_{k=1}^{m} 16\big((c_i - b_i)(c_j - b_j) + (b_i - a_i)(b_j - a_j)\big)(c_k - a_k)x_i x_j x_k \\[2ex]
\qquad \left. + \left| \displaystyle\sum_{i=1}^{m}\sum_{j=1}^{m}\sum_{k=1}^{m} 11(c_i - a_i)(c_j - a_j)(c_k + a_k - 2b_k)x_i x_j x_k \right| \right. \\[2ex]
\qquad\qquad \le \left| \displaystyle\sum_{i=1}^{m} 192\beta(c_i + a_i - 2b_i)x_i \right| + \displaystyle\sum_{i=1}^{m} 192\beta(c_i - a_i)x_i \\[2ex]
\quad \displaystyle\sum_{i=1}^{m}\sum_{j=1}^{m}\sum_{k=1}^{m} (c_i - a_i)(c_j - a_j)(c_k + a_k - 2b_k)x_i x_j x_k \ge \gamma \\[2ex]
\quad x_1 + x_2 + \cdots + x_m = 1 \\[1ex]
\quad x_i \ge 0, \quad i = 1, 2, \ldots, m.
\end{cases}
$$

If $\xi_i = (a_i, b_i)$, $i = 1, 2, \ldots, m$ are mutually independent equipossible fuzzy variables, then model (3.27) has the following crisp equivalent,

$$
\begin{cases}
\max \quad (a_1 + b_1)x_1 + (a_2 + b_2)x_2 + \cdots + (a_m + b_m)x_m \\
\text{s.t.} \quad (b_1 - a_1)x_1 + (b_2 - a_2)x_2 + \cdots + (b_m - a_m)x_m \le \sqrt{8\beta} \\
\qquad x_1 + x_2 + \cdots + x_m = 1 \\
\qquad x_i \ge 0, \quad i = 1, 2, \ldots, m.
\end{cases}
$$

If $\xi_i = (a_i, b_i, c_i, d_i)$, $i = 1, 2, \ldots, m$ are mutually independent symmetric trapezoidal fuzzy variables, model (3.27) has the following crisp equivalent,

$$
\begin{cases}
\max \ \displaystyle\sum_{i=1}^{m} (a_i + b_i + c_i + d_i)x_i \\[2ex]
\text{s.t.} \ \displaystyle\sum_{i=1}^{m}\sum_{j=1}^{m} \big((d_i - a_i)(d_j - a_j) + (d_i - a_i + c_i - b_i)(c_j - b_j)\big)x_i x_j \le 24\beta \\[2ex]
\qquad x_1 + x_2 + \cdots + x_m = 1 \\[1ex]
\qquad x_i \ge 0, \quad i = 1, 2, \ldots, m.
\end{cases}
$$

Table 3.3 Normal fuzzy returns	Stock	Expected value	Standard variance
	1	1.1	1.0
	2	1.3	1.1
	3	1.5	1.3
	4	1.0	1.4

If $\xi_i = EXP(m_i)$, $i = 1, 2, \ldots, m$ are mutually independent exponential fuzzy variables, then model (3.27) has the following crisp equivalent,

$$
\begin{cases}
\max & m_1 x_1 + m_2 x_2 + \cdots + m_m x_m \\
\text{s.t.} & m_1 x_1 + m_2 x_2 + \cdots + m_m x_m \leq \sqrt{\beta/\delta_2} \\
& m_1 x_1 + m_2 x_2 + \cdots + m_m x_m \geq \sqrt[3]{\gamma/\delta_3} \\
& x_1 + x_2 + \cdots + x_m = 1 \\
& x_i \geq 0, \quad i = 1, 2, \ldots, m.
\end{cases}
$$

In order to solve an expected value model, we first check whether there is a crisp equivalent such that we can apply the classical nonlinear programming algorithms to solve the optimal solution. If not, we can solve the suboptimal solution by using the genetic algorithm integrated with the fuzzy simulation, which is used to approximate the expected value operator.

Example 3.21 Suppose that there are four stocks with independent normal fuzzy returns. Table 3.3 shows the expected values and standard variances for these fuzzy quantities. If the investor would like to maximize the investment return with risk level 1.44, then we get the following mean-variance model

$$
\begin{cases}
\max & 1.1 x_1 + 1.3 x_2 + 1.5 x_3 + x_4 \\
\text{s.t.} & x_1 + 1.1 x_2 + 1.3 x_3 + 1.4 x_4 \leq 1.2 \\
& x_1 + x_2 + x_3 + x_4 = 1 \\
& x_1, x_2, x_3, x_4 \geq 0.
\end{cases}
$$

We use the Matlab function *Linprog* to solve the linear programming model. The optimal investment return is 1.3836, and the optimal portfolio is

$$x_1 = 0.1641, \qquad x_2 = 0.2538, \qquad x_3 = 0.5821, \qquad x_4 = 0.0000.$$

Note that the optimal portfolio distributes no capital on the fourth stock since it has the lowest return but has the highest risk.

Example 3.22 In this example, the mean-variance-skewness model (3.27) is applied to the data shown in Table 3.4, which is composed of two triangular fuzzy variables, and two normal fuzzy variables. The genetic algorithm is used to solve the suboptimal solution, which is coded in Matlab programming language under the running

Table 3.4 Fuzzy returns of four stocks

Stock	Fuzzy return
1	$(-0.3, 1.8, 2.3)$
2	$(-0.4, 2.0, 2.2)$
3	$N(1.3, 0.8)$
4	$N(1.5, 1.2)$

environment: a Windows 7 platform of personal computer with processor speed 2.4 GHz and memory size 2 GB.

Assume that the maximum risk level is 1.3 and the minimum skewness level is 0.4. In order to obtain a portfolio which maximizes the investment return, we formulate the following model,

$$\begin{cases} \max & E[\xi_1 x_1 + \xi_2 x_2 + \xi_3 x_3 + \xi_4 x_4] \\ \text{s.t.} & V[\xi_1 x_1 + \xi_2 x_2 + \xi_3 x_3 + \xi_4 x_4] \leq 1.3 \\ & S[\xi_1 x_1 + \xi_2 x_2 + \xi_3 x_3 + \xi_4 x_4] \geq 0.4 \\ & x_1 + x_2 + x_3 + x_4 = 1 \\ & x_1, x_2, x_3, x_4 \geq 0. \end{cases}$$

Take $N = 3000$, $G = 30$, $P_c = 0.4$, $P_m = 0.2$ and *pop-size* $= 100$. A run of the genetic algorithm shows that the best found portfolio is

$$x_1 = 0.0046, \qquad x_2 = 0.0202, \qquad x_3 = 0.5095, \qquad x_4 = 0.4657,$$

and the investment return is 1.3966.

References

Das B, Maity K, Maiti A (2007) A two warehouse supply-chain model under possibility/ necessity/credibility measures. Math Comput Model 46(3–4):398–409

Dubois D, Prade H (1987) The mean value of a fuzzy number. Fuzzy Sets Syst 24:279–300

Feng YQ, Wu WC, Zhang BM, Li WY (2008) Power system operation risk assessment using credibility theory. IEEE Trans Power Syst 23(3):1309–1318

González A (1990) A study of the ranking function approach through mean values. Fuzzy Sets Syst 35:29–41

Heilpern S (1992) The expected value of a fuzzy number. Fuzzy Sets Syst 47:81–86

Ji XY, Iwamura K (2007) New models for shortest path problem with fuzzy arc lengths. Appl Math Model 31:259–269

Ji XY, Shao Z (2006) Model and algorithm for bilevel newsboy problem with fuzzy demands and discounts. Appl Math Comput 172(1):163–174

Ke H, Liu B (2010) Fuzzy project scheduling problem and its hybrid intelligent algorithm. Appl Math Model 34(2):301–308

Li X, Qin ZF, Kar S (2010a) Mean-variance-skewness model for portfolio selection with fuzzy returns. Eur J Oper Res 202:239–247

Li X, Yang L, Gao J (2010c) Fuzzy Edmundson-Madansky inequality and its application to portfolio selection problems. Information 13(4):1163–1173

Liu B (2004) Uncertainty theory: an introduction to its axiomatic foundations. Springer, Berlin

Liu YK (2006) Convergent results about the use of fuzzy simulation in fuzzy optimization problems. IEEE Trans Fuzzy Syst 14(2):295–304

Liu B (2007) Uncertainty theory, 2nd edn. Springer, Berlin

Liu B, Liu YK (2002) Expected value of fuzzy variable and fuzzy expected value models. IEEE Trans Fuzzy Syst 10(4):445–450

Markowitz H (1952) Porfolio selection. J Finance 7:77–91

Peng J, Liu B (2004) Parallel machine scheduling models with fuzzy processing times. Inf Sci 166(1–4):49–66

Samuelson PA (1970) The fundamental approximation theorem of portfolio analysis in terms of means, variances, and higher moments. Rev Econ Stud 37:537–542

Shao Z, Ji XY (2006) Fuzzy multi-product constraint newsboy problem. Appl Math Comput 180(1):7–15

Yang LX, Li KP, Gao ZY (2009) Train timetable problem on a single-line railway with fuzzy passenger demand. IEEE Trans Fuzzy Syst 17(3):617–629

Zhao R, Liu B (2005) Standby redundancy optimization problems with fuzzy lifetimes. Comput Ind Eng 49:318–338

Zhou J, Liu B (2007) Modeling capacitated location-allocation problem with fuzzy demands. Comput Ind Eng 53(3):454–468

Zhu Y, Ji XY (2006) Expected values of functions of fuzzy variables. J Intell Fuzzy Syst 17(5):471–478

Chapter 4
Chance-Constrained Programming Model

Chance-constrained programming (Charnes and Cooper 1961) provides a powerful means of modeling decision systems on the assumption that the constraints will hold at least α of time, where α is the confidence level provided as an approximate safety margin by the decision-maker. For fuzzy decision problems, Liu and Iwamura (1998a,b) introduced a maximax chance-constrained programming model, and Liu (1998) provided a maximin chance-constrained programming model, which respectively maximize the optimistic value and the pessimistic value of the fuzzy objective under certain credibility constraints. Nowadays, fuzzy chance-constrained programming models have been widely used in many real-life applications, such as facility location problem (Zhou and Liu 2007), newsboy problem (Shao and Ji 2006), portfolio selection problem (Li et al. 2010b), project scheduling problem (Ke et al. 2010), quadratic assignment problem (Liu and Li 2006), vehicle routing problem (Zheng and Liu 2006), and so on.

This chapter mainly introduces the concepts of optimistic value and pessimistic value, chance-constrained programming models, fuzzy simulation, and applications in fuzzy portfolio analysis.

4.1 Optimistic Value

Optimistic value and pessimistic value are two important credibilistic mappings which can be used to rank fuzzy variables. This section first introduces the optimistic value.

Definition 4.1 (Liu 2004) Let ξ be a fuzzy variable, and $\alpha \in (0, 1]$. Then

$$\xi_{\sup}(\alpha) = \sup\{r \mid \mathrm{Cr}\{\xi \geq r\} \geq \alpha\} \tag{4.1}$$

is called the α-optimistic value to ξ.

X. Li, *Credibilistic Programming*, Uncertainty and Operations Research,
DOI 10.1007/978-3-642-36376-4_4, © Springer-Verlag Berlin Heidelberg 2013

Fig. 4.1 The α-optimistic value

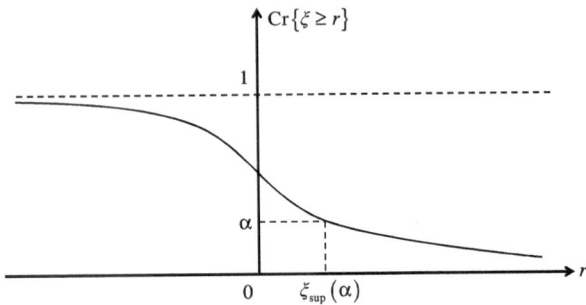

Remark 4.1 The α-optimistic value is the supremum value that the fuzzy variable achieves with credibility α.

Remark 4.2 Denote $\Psi(r) = \text{Cr}\{\xi \geq r\}$. For any $r_1 \leq r_2$, according to the monotonicity axiom of credibility measure, we have

$$\Psi(r_1) \geq \Psi(r_2),$$

which implies that Ψ is a decreasing function (see Fig. 4.1). Furthermore, if function Ψ is strictly decreasing and continuous, then it is clear that the optimistic value function $\xi_{\sup}(\alpha)$ is the inverse function of Ψ.

Example 4.1 Suppose that ξ is a simple fuzzy variable defined by the following credibility function

$$v(x) = \begin{cases} c_1, & \text{if } x = x_1 \\ c_2, & \text{if } x = x_2 \\ \cdots & \cdots \\ c_n, & \text{if } x = x_n. \end{cases}$$

For simplicity, we assume that $x_1 < x_2 < \cdots < x_n$ and $c_1 \leq c_2 \leq \cdots \leq c_n$. It follows from the credibility inversion theorem that

$$\text{Cr}\{\xi \geq r\} = \begin{cases} 1, & \text{if } r \leq x_1 \\ 1 - c_i, & \text{if } x_i < r \leq x_{i+1}, \ 1 \leq i \leq n-1 \\ 0, & \text{if } r > x_n. \end{cases}$$

According to Definition 4.1, the optimistic value is calculated to be

$$\xi_{\sup}(\alpha) = \begin{cases} x_n, & \text{if } 0 < \alpha \leq 1 - c_{n-1} \\ x_i, & \text{if } 1 - c_i < \alpha \leq 1 - c_{i-1}, \ 2 \leq i \leq n-1 \\ x_1, & \text{if } 1 - c_1 < \alpha \leq 1 \end{cases}$$

which is a decreasing and left-continuous function (see Fig. 4.2).

Fig. 4.2 Optimistic value of
a simple fuzzy variable

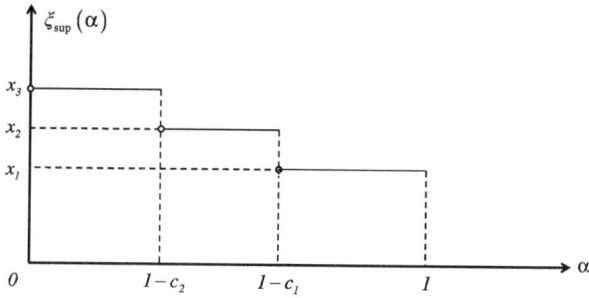

Example 4.2 Let $\xi = (a, b)$ be an equipossible fuzzy variable. It follows from the credibility inversion theorem that

$$\mathrm{Cr}\{\xi \geq r\} = \begin{cases} 1, & \text{if } r \leq a \\ 0.5, & \text{if } a < r \leq b \\ 0, & \text{if } r > b. \end{cases}$$

First, assume $\alpha \leq 0.5$. For any $r \leq b$, it is easy to prove that

$$\mathrm{Cr}\{\xi \geq r\} \geq 0.5 \geq \alpha,$$

and for any $r > b$, we have

$$\mathrm{Cr}\{\xi \geq r\} = 0 < \alpha.$$

Then according to Definition 4.1, we have

$$\xi_{\sup}(\alpha) = \sup\{r \mid \mathrm{Cr}\{\xi \geq r\} \geq \alpha\} = b.$$

Now, assume $\alpha > 0.5$. For any $r \leq a$, it is easy to prove that

$$\mathrm{Cr}\{\xi \geq r\} = 1 \geq \alpha,$$

and for any $r > a$, we have

$$\mathrm{Cr}\{\xi \geq r\} \leq 0.5 < \alpha.$$

Then it follows from Definition 4.1 that $\xi_{\sup}(\alpha) = a$. In general, the α-optimistic value for an equipossible fuzzy variable is

$$\xi_{\sup}(\alpha) = \begin{cases} b, & \text{if } \alpha \leq 0.5 \\ a, & \text{if } \alpha > 0.5 \end{cases}$$

which is shown by Fig. 4.3.

Fig. 4.3 Optimistic value of
an equipossible fuzzy variable

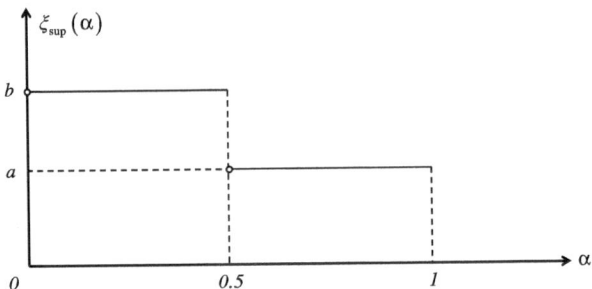

Example 4.3 Let $\xi = (a, b, c)$ be a triangular fuzzy variable. It follows from the
credibility inversion theorem that

$$\mathrm{Cr}\{\xi \geq r\} = \begin{cases} 1, & \text{if } r \leq a \\ (2b - a - r)/2(b - a), & \text{if } a < r \leq b \\ (c - r)/2(c - b), & \text{if } b < r \leq c \\ 0, & \text{if } r > c, \end{cases}$$

which is a strictly decreasing and continuous function on interval $[a, c]$. For any
$\alpha \leq 0.5$, it is easy to prove that

$$\xi_{\sup}(\alpha) = \sup\{b < r \leq c \mid (c - r)/2(c - b) \geq \alpha\} = 2\alpha b + (1 - 2\alpha)c.$$

Similarly, for any $\alpha > 0.5$, we have

$$\xi_{\sup}(\alpha) = \sup\{a < r \leq b \mid (2b - a - r)/2(b - a) \geq \alpha\}$$
$$= (2\alpha - 1)a + (2 - 2\alpha)b.$$

In general, the α-optimistic value for a triangular fuzzy variable is

$$\xi_{\sup}(\alpha) = \begin{cases} 2\alpha b + (1 - 2\alpha)c, & \text{if } \alpha \leq 0.5 \\ (2\alpha - 1)a + (2 - 2\alpha)b, & \text{if } \alpha > 0.5 \end{cases}$$

which is shown by Fig. 4.4.

Example 4.4 Suppose that $\xi = (a, b, c, d)$ is a trapezoidal fuzzy variable. It follows
from the credibility inversion theorem that

$$\mathrm{Cr}\{\xi \geq r\} = \begin{cases} 1, & \text{if } r \leq a \\ (2b - a - r)/2(b - a), & \text{if } a < r \leq b \\ 0.5, & \text{if } b < r \leq c \\ (d - r)/2(d - c), & \text{if } c < r \leq d \\ 0, & \text{if } r > d. \end{cases}$$

Fig. 4.4 Optimistic value of
a triangular fuzzy variable

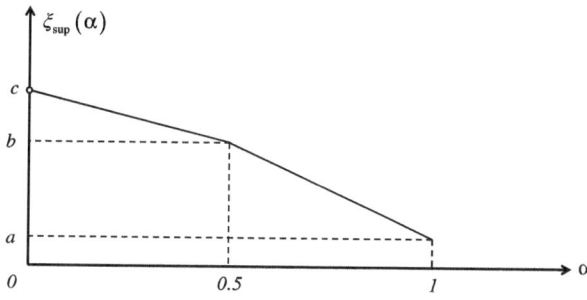

Fig. 4.5 Optimistic value of
a trapezoidal fuzzy variable

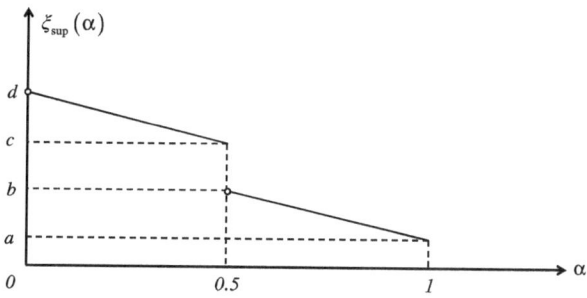

For any $\alpha < 0.5$, it is easy to prove that

$$\xi_{\sup}(\alpha) = \sup\{c < r \leq d \mid (d - r)/2(d - c) \geq \alpha\}$$
$$= 2\alpha c + (1 - 2\alpha)d.$$

Similarly, for any $\alpha > 0.5$, we have

$$\xi_{\sup}(\alpha) = \sup\{a < r \leq b \mid (2b - a - r)/2(b - a) \geq \alpha\}$$
$$= (2\alpha - 1)a + (2 - 2\alpha)b.$$

Especially, when $\alpha = 0.5$, we have $\xi_{\sup}(\alpha) = c$. In general, the α-optimistic value for a trapezoidal fuzzy variable is

$$\xi_{\sup}(\alpha) = \begin{cases} 2\alpha c + (1 - 2\alpha)d, & \text{if } \alpha \leq 0.5 \\ (2\alpha - 1)a + (2 - 2\alpha)b, & \text{if } \alpha > 0.5 \end{cases}$$

which is shown by Fig. 4.5.

Example 4.5 Let $\xi = EXP(m)$ be an exponential fuzzy variable. Then it follows from the credibility inversion theorem that

$$\text{Cr}\{\xi \geq r\} = \begin{cases} 1, & \text{if } r \leq 0 \\ 1/(1 + \exp(\pi r/\sqrt{6}m)), & \text{if } r > 0. \end{cases}$$

Fig. 4.6 Optimistic value of an exponential fuzzy variable

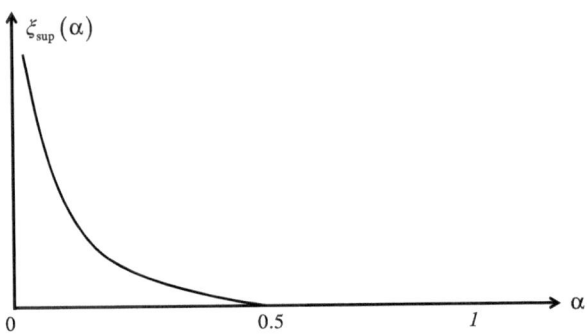

According to Definition 4.1, for any $\alpha < 0.5$, we have

$$\xi_{\sup}(\alpha) = \sup\{r \mid 1/(1 + \exp(\pi r/\sqrt{6}m)) \geq \alpha\}$$
$$= (\ln(1 - \alpha) - \ln \alpha)\sqrt{6}m/\pi.$$

Otherwise, when $\alpha \geq 0.5$, we have $\xi_{\sup}(\alpha) = 0$. In general, the optimistic value function for an exponential fuzzy variable is

$$\xi_{\sup}(\alpha) = \max\{(\ln(1 - \alpha) - \ln \alpha)\sqrt{6}m/\pi, 0\}$$

which is shown by Fig. 4.6.

Example 4.6 Suppose that ξ is a normal fuzzy variable $N(e, \sigma)$. It follows from the credibility inversion theorem that

$$\mathrm{Cr}\{\xi \geq r\} = 1/(1 + \exp(\pi(r - e)/\sqrt{6}\sigma))$$

which is a strictly decreasing and continuous function. For any $\alpha \in (0, 1]$, according to Definition 4.1, we have

$$\xi_{\sup}(\alpha) = \sup\{r \mid 1/(1 + \exp(\pi(r - e)/\sqrt{6}\sigma)) \geq \alpha\}$$
$$= e + (\ln(1 - \alpha) - \ln \alpha)\sqrt{6}\sigma/\pi$$

which is shown by Fig. 4.7.

Theorem 4.1 (Liu 2004) *The α-optimistic value is a decreasing and left-continuous function with respect to α.*

Proof It is easy to prove that $\xi_{\sup}(\alpha)$ is a decreasing function of α. Next, we prove the left-continuity. Let α_i be an arbitrary sequence of positive numbers such that $\alpha_i \uparrow \alpha$. Then $\{\xi_{\sup}(\alpha_i)\}$ is a decreasing sequence. If the limitation is equal to $\xi_{\sup}(\alpha)$, then the left-continuity is proved. Otherwise, there exists a number z such that

$$\lim_{i \to \infty} \xi_{\sup}(\alpha_i) > z > \xi_{\sup}(\alpha).$$

Fig. 4.7 Optimistic value of
a normal fuzzy variable

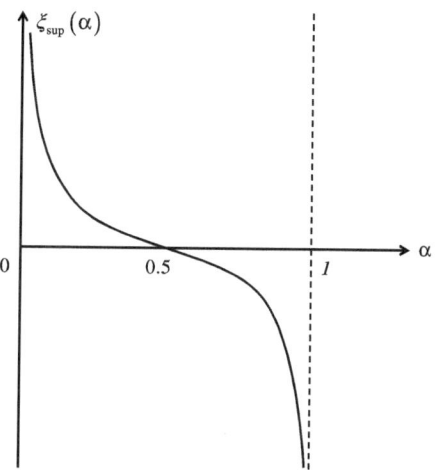

According to the definition of optimistic value, we have $\mathrm{Cr}\{\xi \geq z\} \geq \alpha_i$ for each α_i. Letting $i \to \infty$, we get $\mathrm{Cr}\{\xi \geq z\} \geq \alpha$. Hence $z \leq \xi_{\sup}(\alpha)$. The contradiction proves the left-continuity. The proof is complete. □

Theorem 4.2 (Liu 2004) *Suppose that $\xi_{\sup}(\alpha)$ is the α-optimistic value of fuzzy variable ξ. For any $\alpha > 0.5$, we have*

$$\mathrm{Cr}\big\{\xi \geq \xi_{\sup}(\alpha)\big\} \geq \alpha. \tag{4.2}$$

Proof For any $\alpha > 0.5$, it follows from the definition of α-optimistic value that there exists an increasing sequence $\{x_i\}$ with limit $\xi_{\sup}(\alpha)$ such that $\mathrm{Cr}\{\xi \geq x_i\} \geq \alpha$ for all $i = 1, 2, \ldots$. Since $\{\xi \geq x_i\} \downarrow \{\xi \geq \xi_{\sup}(\alpha)\}$ and

$$\lim_{i \to \infty} \mathrm{Cr}\{\xi \geq x_i\} > 0.5,$$

it follows from the credibility semicontinuous theorem that

$$\mathrm{Cr}\big\{\xi \geq \xi_{\sup}(\alpha)\big\} = \lim_{i \to \infty} \mathrm{Cr}\{\xi \geq x_i\} \geq \alpha.$$

The proof is complete. □

Remark 4.3 When $\alpha \leq 0.5$, it is possible that $\mathrm{Cr}\{\xi \geq \xi_{\sup}(\alpha)\} < \alpha$. For example, let ξ be a fuzzy variable with credibility function

$$\nu(x) = 0.5, \quad \forall x \in (0, 1).$$

For any $0 < \alpha \leq 0.5$, it is easy to prove that $\xi_{\sup}(\alpha) = 1$. However, we have

$$\mathrm{Cr}\{\xi \geq 1\} = 0 < \alpha.$$

Theorem 4.3 *If ξ is a continuous fuzzy variable, then we have*

$$\mathrm{Cr}\big\{\xi \geq \xi_{\sup}(\alpha)\big\} \geq \alpha. \tag{4.3}$$

Proof For any $\alpha \in (0, 1]$, according to Definition 4.1, there exists an increasing sequence $\{x_i\}$ with limit $\xi_{\sup}(\alpha)$ such that $\mathrm{Cr}\{\xi \geq x_i\} \geq \alpha$ for all $i = 1, 2, \ldots$. Since ξ has a continuous credibility function, it is easy to prove that the function $\mathrm{Cr}\{\xi \geq r\}$ is also continuous with respect to r. Then it follows from the continuity that

$$\mathrm{Cr}\big\{\xi \geq \xi_{\sup}(\alpha)\big\} = \lim_{i \to \infty} \mathrm{Cr}\{\xi \geq x_i\} \geq \alpha.$$

The proof is complete.   □

Theorem 4.4 (Li and Liu 2006b) *Suppose that fuzzy variables ξ and η are mutually independent. Then for any $\alpha \in (0, 1]$, we have*

$$(\xi + \eta)_{\sup}(\alpha) = \xi_{\sup}(\alpha) + \eta_{\sup}(\alpha). \tag{4.4}$$

Proof For any given number $\varepsilon > 0$, it is easy to prove the following relations

$$\big\{\xi + \eta \geq \xi_{\sup}(\alpha) + \eta_{\sup}(\alpha) - \varepsilon\big\} \supseteq \big\{\xi \geq \xi_{\sup}(\alpha) - \varepsilon/2\big\} \cap \big\{\eta \geq \eta_{\sup}(\alpha) - \varepsilon/2\big\},$$

$$\big\{\xi + \eta \geq \xi_{\sup}(\alpha) + \eta_{\sup}(\alpha) + \varepsilon\big\} \subseteq \big\{\xi \geq \xi_{\sup}(\alpha) + \varepsilon/2\big\} \cup \big\{\eta \geq \eta_{\sup}(\alpha) + \varepsilon/2\big\}.$$

According to the definition of optimistic value, for any $\alpha \in (0, 1]$, we have

$$\mathrm{Cr}\big\{\xi \geq \xi_{\sup}(\alpha) - \varepsilon/2\big\} \geq \alpha > \mathrm{Cr}\big\{\xi \geq \xi_{\sup}(\alpha) + \varepsilon/2\big\},$$

$$\mathrm{Cr}\big\{\eta \geq \eta_{\sup}(\alpha) - \varepsilon/2\big\} \geq \alpha > \mathrm{Cr}\big\{\eta \geq \eta_{\sup}(\alpha) + \varepsilon/2\big\}.$$

Since fuzzy variables ξ and η are mutually independent, it follows from the monotonicity axiom that

$$\mathrm{Cr}\big\{\xi + \eta \geq \xi_{\sup}(\alpha) + \eta_{\sup}(\alpha) - \varepsilon\big\} \geq \alpha > \mathrm{Cr}\big\{\xi + \eta \geq \xi_{\sup}(\alpha) + \eta_{\sup}(\alpha) + \varepsilon\big\}$$

which implies that $\xi_{\sup}(\alpha) + \eta_{\sup}(\alpha) + \varepsilon \geq (\xi + \eta)_{\sup}(\alpha) \geq \xi_{\sup}(\alpha) + \eta_{\sup}(\alpha) - \varepsilon$. Letting $\varepsilon \to 0$, we obtain equation (4.4). The proof is complete.   □

Example 4.7 The independence assumption cannot be removed in above theorem. For example, take a credibility space $(\Theta, \mathcal{A}, \mathrm{Cr})$ to be $\{\theta_1, \theta_2\}$ with $\mathrm{Cr}\{\theta_1\} = \mathrm{Cr}\{\theta_2\} = 0.5$. Define fuzzy variables ξ and η as follows,

$$\xi(\theta) = \begin{cases} 2, & \text{if } \theta = \theta_1 \\ 1, & \text{if } \theta = \theta_2, \end{cases} \qquad \eta(\theta) = \begin{cases} 1, & \text{if } \theta = \theta_1 \\ 2, & \text{if } \theta = \theta_2. \end{cases}$$

Taking $\alpha = 0.6$, it is easy to prove that $\xi_{\sup}(\alpha) = \eta_{\sup}(\alpha) = 1$ and $(\xi + \eta)_{\sup}(\alpha) = 3$, which implies that

$$(\xi + \eta)_{\sup}(\alpha) > \xi_{\sup}(\alpha) + \eta_{\sup}(\alpha).$$

Theorem 4.5 (Li and Liu 2006b) *Suppose that ξ and η are two independent non-negative fuzzy variables. Then for any $\alpha \in (0, 1]$, we have*

$$(\xi\eta)_{\sup}(\alpha) = \xi_{\sup}(\alpha)\eta_{\sup}(\alpha). \tag{4.5}$$

Proof For any given real number $\varepsilon > 0$, since fuzzy variables ξ and η are both nonnegative, it is easy to prove that

$$\{\xi\eta \geq (\xi_{\sup}(\alpha) + \varepsilon)(\eta_{\sup}(\alpha) + \varepsilon)\} \subseteq \{\xi \geq \xi_{\sup}(\alpha) + \varepsilon\} \cup \{\eta \geq \eta_{\sup}(\alpha) + \varepsilon\}.$$

It follows from the monotonicity axiom and the independence that

$$\text{Cr}\{\xi\eta \geq (\xi_{\sup}(\alpha) + \varepsilon)(\eta_{\sup}(\alpha) + \varepsilon)\}$$
$$\leq \text{Cr}\{\xi \geq \xi_{\sup}(\alpha) + \varepsilon\} \vee \text{Cr}\{\eta \geq \eta_{\sup}(\alpha) + \varepsilon\}$$
$$< \alpha.$$

Then according to the definition of optimistic value, we have

$$(\xi\eta)_{\sup}(\alpha) \leq (\xi_{\sup}(\alpha) + \varepsilon)(\eta_{\sup}(\alpha) + \varepsilon).$$

When $\xi_{\sup}(\alpha)\eta_{\sup}(\alpha) = 0$, letting $\varepsilon \to 0$, we obtain $(\xi\eta)_{\sup}(\alpha) = 0$. Otherwise, let ε_0 be a positive number with $\varepsilon_0 < \varepsilon$ such that

$$\xi_{\sup}(\alpha) - \varepsilon_0 > 0, \qquad \eta_{\sup}(\alpha) - \varepsilon_0 > 0.$$

Based on the relation

$$\{\xi\eta \geq (\xi_{\sup}(\alpha) - \varepsilon_0)(\eta_{\sup}(\alpha) - \varepsilon_0)\} \supseteq \{\xi \geq \xi_{\sup}(\alpha) - \varepsilon_0\} \cap \{\eta \geq \eta_{\sup}(\alpha) - \varepsilon_0\},$$

it follows from the monotonicity axiom and the independence that

$$\text{Cr}\{\xi\eta \geq (\xi_{\sup}(\alpha) - \varepsilon_0)(\eta_{\sup}(\alpha) - \varepsilon_0)\}$$
$$\geq \text{Cr}\{\xi \geq \xi_{\sup}(\alpha) - \varepsilon_0\} \wedge \text{Cr}\{\eta \geq \eta_{\sup}(\alpha) - \varepsilon_0\}$$
$$\geq \alpha$$

which implies that $(\xi\eta)_{\sup}(\alpha) \geq (\xi_{\sup}(\alpha) - \varepsilon_0)(\eta_{\sup}(\alpha) - \varepsilon_0)$. In general, we get

$$(\xi_{\sup}(\alpha) - \varepsilon_0)(\eta_{\sup}(\alpha) - \varepsilon_0) \leq (\xi\eta)_{\sup}(\alpha) \leq (\xi_{\sup}(\alpha) + \varepsilon)(\eta_{\sup}(\alpha) + \varepsilon).$$

Letting $\varepsilon \to 0$, we obtain (4.5). The proof is complete.   □

Example 4.8 Similarly, the independence condition cannot be removed in Theorem 4.5. Let us reconsider Example 4.7 It is easy to prove that

$$(\xi\eta)_{\sup}(0.6) = 2 > 1 = \xi_{\sup}(0.6)\eta_{\sup}(0.6).$$

Theorem 4.6 *Suppose that fuzzy variables ξ and η are mutually independent. Then for any $\alpha \in (0, 1]$, we have*

$$(\xi \wedge \eta)_{\sup}(\alpha) = \xi_{\sup}(\alpha) \wedge \eta_{\sup}(\alpha). \tag{4.6}$$

Proof Like Theorem 4.5 except that the "$\times$" is replaced with "$\wedge$".   □

Theorem 4.7 *Suppose that fuzzy variables ξ and η are mutually independent. Then for any $\alpha \in (0, 1]$, we have*

$$(\xi \vee \eta)_{\sup}(\alpha) = \xi_{\sup}(\alpha) \vee \eta_{\sup}(\alpha). \tag{4.7}$$

Proof Like Theorem 4.5 except that the "$\times$" is replaced with "$\vee$".   □

4.2 Pessimistic Value

This section introduces the concept of pessimistic value, which has some similar properties with the optimistic value.

Definition 4.2 (Liu 2004) Let ξ be a fuzzy variable, and $\alpha \in (0, 1]$. Then

$$\xi_{\inf}(\alpha) = \inf\{r \mid \mathrm{Cr}\{\xi \leq r\} \geq \alpha\} \tag{4.8}$$

is called the α-pessimistic value to ξ.

Remark 4.4 The α-pessimistic value is the infimum value that the fuzzy variable achieves with credibility α.

Remark 4.5 Denote $\Phi(r) = \mathrm{Cr}\{\xi \leq r\}$. For any $r_1 \leq r_2$, according to the monotonicity axiom of credibility measure, we have

$$\Phi(r_1) \leq \Phi(r_2)$$

which implies that $\Phi(r)$ is an increasing function with respect to r. See Fig. 4.8. Furthermore, if it is strictly increasing and continuous, then it is clear that $\xi_{\inf}(\alpha)$ is the inverse function of Φ.

Example 4.9 Suppose that ξ is a simple fuzzy variable defined by the following credibility function

$$\nu(x) = \begin{cases} c_1, & \text{if } x = x_1 \\ c_2, & \text{if } x = x_2 \\ \cdots & \cdots \\ c_n, & \text{if } x = x_n. \end{cases}$$

Fig. 4.8 The α-pessimistic value

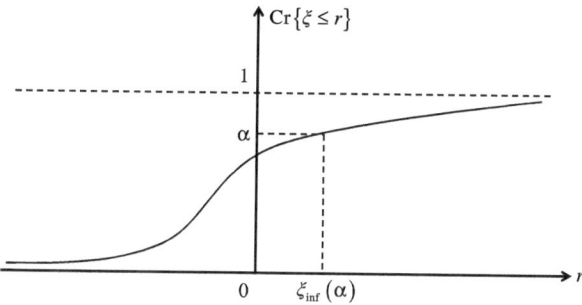

Fig. 4.9 Pessimistic value of a simple fuzzy variable

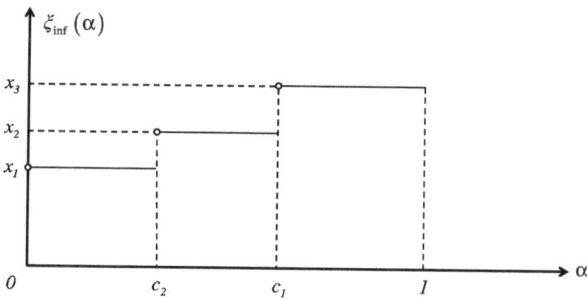

For simplicity, we assume that $x_1 < x_2 < \cdots < x_n$ and $c_1 \leq c_2 \leq \cdots \leq c_n$. It follows from the credibility inversion theorem that

$$\mathrm{Cr}\{\xi \leq r\} = \begin{cases} 0, & \text{if } r < x_1 \\ c_i, & \text{if } x_i \leq r < x_{i+1}, \ 1 \leq i \leq n-1 \\ 1, & \text{if } r \geq x_n. \end{cases}$$

According to Definition 4.2, the pessimistic value is calculated to be

$$\xi_{\inf}(\alpha) = \begin{cases} x_1, & \text{if } 0 < \alpha \leq c_1 \\ x_i, & \text{if } c_i < \alpha \leq c_{i+1}, \ 1 \leq i \leq n-1 \\ x_n, & \text{if } c_n < \alpha \leq 1 \end{cases}$$

which is an increasing and left-continuous function. See Fig. 4.9.

Example 4.10 Let $\xi = (a, b)$ be an equipossible fuzzy variable. It follows from the credibility inversion theorem that

$$\mathrm{Cr}\{\xi \leq r\} = \begin{cases} 0, & \text{if } r < a \\ 0.5, & \text{if } a \leq r < b \\ 1, & \text{if } r \geq b. \end{cases}$$

Fig. 4.10 Pessimistic value
of an equipossible fuzzy
variable

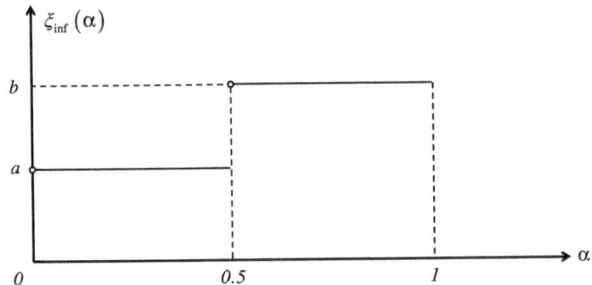

First, we assume $\alpha \leq 0.5$. For any $r \geq a$, it is easy to prove that

$$\mathrm{Cr}\{\xi \leq r\} \geq 0.5 \geq \alpha,$$

and for any $r < a$, we have

$$\mathrm{Cr}\{\xi \leq r\} = 0 \leq \alpha.$$

Then according to Definition 4.2, we have

$$\xi_{\inf}(\alpha) = \inf\{r \mid \mathrm{Cr}\{\xi \leq r\} \geq \alpha\} = a.$$

Similarly, for any $\alpha > 0.5$, we can prove $\xi_{\inf}(\alpha) = b$. In general, the α-pessimistic
value for an equipossible fuzzy variable is

$$\xi_{\inf}(\alpha) = \begin{cases} a, & \text{if } \alpha \leq 0.5 \\ b, & \text{if } \alpha > 0.5 \end{cases}$$

which is shown by Fig. 4.10.

Example 4.11 Let $\xi = (a, b, c)$ be a triangular fuzzy variable. It follows from the
credibility inversion theorem that

$$\mathrm{Cr}\{\xi \leq r\} = \begin{cases} 0, & \text{if } r < a \\ (r-a)/2(b-a), & \text{if } a \leq r < b \\ (c-2b+r)/2(c-b), & \text{if } b \leq r < c \\ 1, & \text{if } r \geq c, \end{cases}$$

which is a strictly increasing and continuous function on interval $[a, c]$. For any
$\alpha \leq 0.5$, it is easy to prove that

$$\xi_{\inf}(\alpha) = \inf\{a \leq r \leq b \mid (r-a)/2(b-a) \geq \alpha\} = 2\alpha b + (1 - 2\alpha)a.$$

Similarly, for any $\alpha > 0.5$, we have

$$\xi_{\inf}(\alpha) = \inf\{b \leq r \leq c \mid (c-2b+r)/2(c-b) \geq \alpha\}$$
$$= (2\alpha - 1)c + (2 - 2\alpha)b.$$

Fig. 4.11 Pessimistic value
of a triangular fuzzy variable

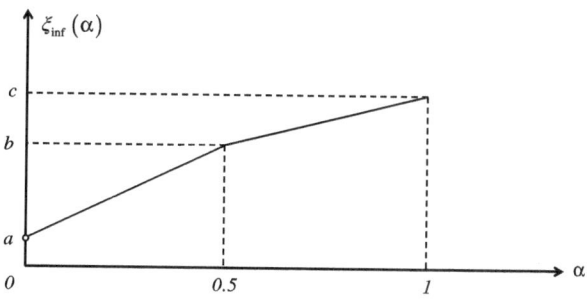

In general, the α-pessimistic value for a triangular fuzzy variable is

$$\xi_{\inf}(\alpha) = \begin{cases} 2\alpha b + (1 - 2\alpha)a, & \text{if } \alpha \leq 0.5 \\ (2\alpha - 1)c + (2 - 2\alpha)b, & \text{if } \alpha > 0.5 \end{cases}$$

which is shown by Fig. 4.11.

Example 4.12 Let $\xi = (a, b, c, d)$ be a trapezoidal fuzzy variable. It follows from the credibility inversion theorem that

$$\text{Cr}\{\xi \leq r\} = \begin{cases} 0, & \text{if } r < a \\ (r - a)/2(b - a), & \text{if } a \leq r < b \\ 0.5, & \text{if } b \leq r < c \\ (d - 2c + r)/2(d - c), & \text{if } c \leq r < d \\ 1, & \text{if } r \geq d. \end{cases}$$

First, for any $\alpha \leq 0.5$, it is easy to prove that

$$\xi_{\inf}(\alpha) = \inf\{a \leq r \leq b \mid (r - a)/2(b - a) \geq \alpha\} = 2\alpha b + (1 - 2\alpha)a.$$

Similarly, for any $\alpha > 0.5$, we have

$$\xi_{\inf}(\alpha) = \inf\{c \leq r \leq d \mid (d - 2c + r)/2(d - c) \geq \alpha\}$$
$$= (2\alpha - 1)d + (2 - 2\alpha)c.$$

In general, the pessimistic value function for a trapezoidal fuzzy variable is

$$\xi_{\inf}(\alpha) = \begin{cases} 2\alpha b + (1 - 2\alpha)a, & \text{if } \alpha \leq 0.5 \\ (2\alpha - 1)d + (2 - 2\alpha)c, & \text{if } \alpha > 0.5 \end{cases}$$

which is shown by Fig. 4.12.

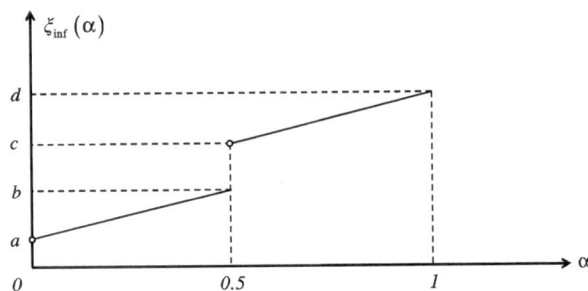

Fig. 4.12 Pessimistic value of a trapezoidal fuzzy variable

Example 4.13 Let $\xi = EXP(m)$ be an exponential fuzzy variable. It follows from the credibility inversion theorem that

$$\text{Cr}\{\xi \leq r\} = \begin{cases} 0, & \text{if } r < 0 \\ 1/(1 + \exp(-\pi r/\sqrt{6}m)), & \text{if } r \geq 0. \end{cases}$$

For any $\alpha > 0.5$, according to Definition 4.2, we have

$$\xi_{\text{inf}}(\alpha) = \inf\{r \mid 1/(1 + \exp(-\pi r/\sqrt{6}m)) \geq \alpha\}$$
$$= (\ln \alpha - \ln(1 - \alpha))\sqrt{6}m/\pi.$$

Otherwise, we have $\xi_{\text{inf}}(\alpha) = 0$. In general, the pessimistic value function for an exponential fuzzy variable is

$$\xi_{\text{inf}}(\alpha) = \max\{(\ln \alpha - \ln(1 - \alpha))\sqrt{6}m/\pi, 0\}$$

which is shown by Fig. 4.13.

Example 4.14 Let $\xi = N(e, \sigma)$ be a normal fuzzy variable. For any $r \in \Re$, it follows from the credibility inversion theorem that

$$\text{Cr}\{\xi \leq r\} = 1/(1 + \exp(\pi(e - r)/\sqrt{6}\sigma)).$$

Then for any $\alpha \in (0, 1]$, according to Definition 4.2, we have

$$\xi_{\text{inf}}(\alpha) = \inf\{r \mid 1/(1 + \exp(\pi(e - r)/\sqrt{6}\sigma)) \geq \alpha\}$$
$$= e - (\ln(1 - \alpha) - \ln \alpha)\sqrt{6}\sigma/\pi$$

which is shown by Fig. 4.14.

Theorem 4.8 (Liu 2004) *The pessimistic value $\xi_{\text{inf}}(\alpha)$ is an increasing and left-continuous function of α.*

Fig. 4.13 Pessimistic value
of an exponential fuzzy
variable

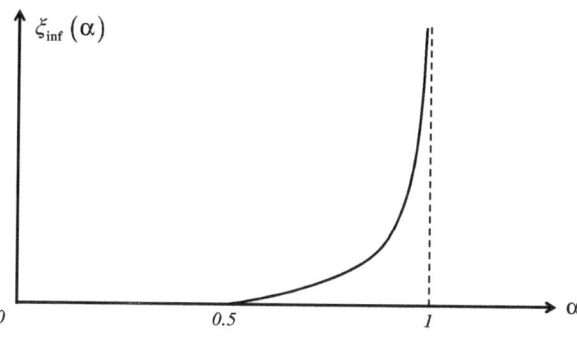

Fig. 4.14 Pessimistic value
of a normal fuzzy variable

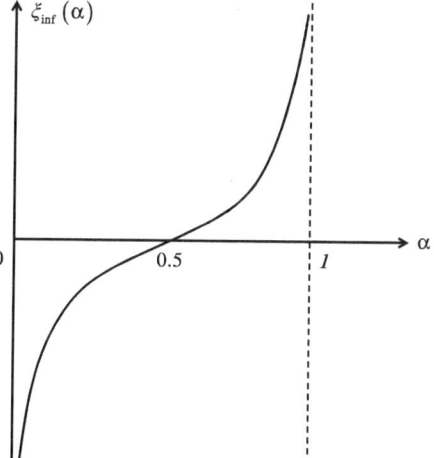

Proof It is easy to prove that $\xi_{\inf}(\alpha)$ is an increasing function with respect to α.
Next, we prove the left-continuity. Let α_i be an arbitrary sequence of positive num-
bers such that $\alpha_i \uparrow \alpha$. Then $\{\xi_{\inf}(\alpha_i)\}$ is an increasing sequence. If the limitation is
equal to $\xi_{\inf}(\alpha)$, then the left-continuity is proved. Otherwise, there exists a number z such that

$$\lim_{i \to \infty} \xi_{\inf}(\alpha) < z < \xi_{\inf}(\alpha).$$

According to Definition 4.2, we have $\mathrm{Cr}\{\xi \le z\} \ge \alpha_i$. Letting $i \to \infty$, we get $\mathrm{Cr}\{\xi \le z\} \ge \alpha$, which implies that $z \ge \xi_{\inf}(\alpha)$. The contradiction proves the left-continuity.
The proof is complete. □

Theorem 4.9 (Liu 2004) *Let ξ be a fuzzy variable. If $\alpha > 0.5$, then we have*

$$\mathrm{Cr}\{\xi \le \xi_{\inf}(\alpha)\} \ge \alpha. \tag{4.9}$$

Proof It follows from the definition of α-pessimistic value that there exists a de-
creasing sequence $\{x_i\}$ with limit $\xi_{\inf}(\alpha)$ such that $\mathrm{Cr}\{\xi \le x_i\} \ge \alpha$ for all $i = 1, 2, \ldots$. Since $\{\xi \le x_i\} \downarrow \{\xi \le \xi_{\inf}(\alpha)\}$ and

$$\lim_{i\to\infty} \text{Cr}\{\xi \le x_i\} > 0.5,$$

it follows from the credibility semicontinuous theorem that

$$\text{Cr}\{\xi \le \xi_{\text{inf}}(\alpha)\} = \lim_{i\to\infty} \text{Cr}\{\xi \le x_i\} \ge \alpha.$$

The proof is complete.   □

Remark 4.6 When $\alpha \le 0.5$, it is possible that (4.9) does not hold. For example, let ξ be a fuzzy variable defined by credibility function

$$\nu(x) = 0.5, \quad \forall x \in (0, 1).$$

For any $0 < \alpha \le 0.5$, it is easy to prove that $\xi_{\text{inf}}(\alpha) = 0$. However, we have

$$\text{Cr}\{\xi \le 0\} = 0 < \alpha.$$

Theorem 4.10 *If ξ is a continuous fuzzy variable, then we have*

$$\text{Cr}\{\xi \le \xi_{\text{inf}}(\alpha)\} \ge \alpha. \tag{4.10}$$

Proof For any $\alpha \in (0, 1]$, it follows from Definition 4.2 that there exists a decreasing sequence $\{x_i\}$ with limit $\xi_{\text{inf}}(\alpha)$ such that $\text{Cr}\{\xi \le x_i\} \ge \alpha$ for all $i = 1, 2, \ldots$. Since ξ has a continuous credibility function, it is easy to prove that $\text{Cr}\{\xi \le r\}$ is also continuous with respect to r, which implies that

$$\text{Cr}\{\xi \le \xi_{\text{inf}}(\alpha)\} = \lim_{i\to\infty} \text{Cr}\{\xi \le x_i\} \ge \alpha.$$

The proof is complete.   □

Theorem 4.11 (Liu 2004) *Suppose that ξ is a fuzzy variable. Then*

(a) *if $\alpha > 0.5$, we have $\xi_{\text{inf}}(\alpha) \ge \xi_{\text{sup}}(\alpha)$;*
(b) *if $\alpha \le 0.5$, we have $\xi_{\text{inf}}(\alpha) \le \xi_{\text{sup}}(\alpha)$.*

Proof Write $z = (\xi_{\text{inf}}(\alpha) + \xi_{\text{sup}}(\alpha))/2$. (a) If $\xi_{\text{inf}}(\alpha) < \xi_{\text{sup}}(\alpha)$, we have

$$1 \ge \text{Cr}\{\xi < z\} + \text{Cr}\{\xi > z\} \ge \alpha + \alpha > 1.$$

The contradiction proves that $\xi_{\text{inf}}(\alpha) \ge \xi_{\text{sup}}(\alpha)$.
    (b) If $\xi_{\text{inf}}(\alpha) > \xi_{\text{sup}}(\alpha)$, it follows from the duality axiom that

$$1 \le \text{Cr}\{\xi \le z\} + \text{Cr}\{\xi \ge z\} < \alpha + \alpha \le 1.$$

The contradiction proves that $\xi_{\text{inf}}(\alpha) \le \xi_{\text{sup}}(\alpha)$. The proof is complete.   □

Theorem 4.12 (Liu 2004) *Suppose that ξ is a fuzzy variable. Then*

(a) *if $c \geq 0$, we have $(c\xi)_{\text{sup}}(\alpha) = c\xi_{\text{sup}}(\alpha)$ and $(c\xi)_{\text{inf}}(\alpha) = c\xi_{\text{inf}}(\alpha)$;*
(b) *if $c < 0$, we have $(c\xi)_{\text{sup}}(\alpha) = c\xi_{\text{inf}}(\alpha)$ and $(c\xi)_{\text{inf}}(\alpha) = c\xi_{\text{sup}}(\alpha)$.*

Proof (a) If $c = 0$, the conclusions obviously hold. When $c > 0$, we have

$$(c\xi)_{\text{sup}}(\alpha) = \sup\{r \mid \text{Cr}\{c\xi \geq r\} \geq \alpha\}$$
$$= c \sup\{r/c \mid \text{Cr}\{\xi \geq r/c\} \geq \alpha\}$$
$$= c\xi_{\text{sup}}(\alpha).$$

A similar way may prove that $(c\xi)_{\text{inf}}(\alpha) = c\xi_{\text{inf}}(\alpha)$.

(b) In order to prove (b), it is sufficient to prove that $(-\xi)_{\text{sup}}(\alpha) = -\xi_{\text{inf}}(\alpha)$ and $(-\xi)_{\text{inf}}(\alpha) = -\xi_{\text{sup}}(\alpha)$. In fact, for any $\alpha \in (0, 1]$, we have

$$(-\xi)_{\text{sup}}(\alpha) = \sup\{r \mid \text{Cr}\{-\xi \geq r\} \geq \alpha\}$$
$$= -\inf\{-r \mid \text{Cr}\{\xi \leq -r\} \geq \alpha\}$$
$$= -\xi_{\text{inf}}(\alpha).$$

A similar way may prove that $(-\xi)_{\text{inf}}(\alpha) = -\xi_{\text{sup}}(\alpha)$. The proof is complete. □

Theorem 4.13 (Li and Liu 2006b) *Suppose that ξ and η are mutually independent fuzzy variables. Then for any $\alpha \in (0, 1]$, we have*

$$(\xi + \eta)_{\text{inf}}(\alpha) = \xi_{\text{inf}}(\alpha) + \eta_{\text{inf}}(\alpha). \tag{4.11}$$

Proof According to Theorems 4.4 and 4.12, it is easy to prove that

$$(\xi + \eta)_{\text{inf}}(\alpha) = -(-\xi - \eta)_{\text{sup}}(\alpha) = -(-\xi)_{\text{sup}}(\alpha) - (-\eta)_{\text{sup}}(\alpha)$$
$$= \xi_{\text{inf}}(\alpha) + \eta_{\text{inf}}(\alpha).$$

The proof is complete. □

Example 4.15 Note that the independence condition cannot be removed in Theorem 4.13. In Example 4.7, it is easy to prove that $\xi_{\text{inf}}(0.6) = \eta_{\text{inf}}(0.6) = 2$ and $(\xi + \eta)_{\text{inf}}(0.6) = 3$, which implies that

$$(\xi + \eta)_{\text{inf}}(0.6) < \xi_{\text{inf}}(0.6) + \eta_{\text{inf}}(0.6).$$

Theorem 4.14 (Li and Liu 2006b) *Suppose that ξ and η are independent fuzzy variables. Then we have*

(a) $(\xi\eta)_{\text{inf}}(\alpha) = \xi_{\text{inf}}(\alpha)\eta_{\text{inf}}(\alpha)$, *if $\xi \geq 0$ and $\eta \geq 0$;*
(b) $(\xi\eta)_{\text{inf}}(\alpha) = \xi_{\text{sup}}(\alpha)\eta_{\text{sup}}(\alpha)$, *if $\xi \leq 0$ and $\eta \leq 0$;*
(c) $(\xi\eta)_{\text{inf}}(\alpha) = \xi_{\text{sup}}(\alpha)\eta_{\text{inf}}(\alpha)$, *if $\xi \geq 0$ and $\eta \leq 0$.*

Proof (a) For any given number $\varepsilon > 0$, since fuzzy variables ξ and η are both non-negative, it is easy to prove that

$$\left\{\xi\eta \leq \left(\xi_{\inf}(\alpha) + \varepsilon\right)\left(\eta_{\inf}(\alpha) + \varepsilon\right)\right\} \supseteq \left\{\xi \leq \xi_{\inf}(\alpha) + \varepsilon\right\} \cap \left\{\eta \leq \eta_{\inf}(\alpha) + \varepsilon\right\}.$$

Then it follows from the independence and the monotonicity axiom that

$$\mathrm{Cr}\left\{\xi\eta \leq \left(\xi_{\inf}(\alpha) + \varepsilon\right)\left(\eta_{\inf}(\alpha) + \varepsilon\right)\right\}$$
$$\geq \mathrm{Cr}\left\{\xi \leq \xi_{\inf}(\alpha) + \varepsilon\right\} \wedge \mathrm{Cr}\left\{\eta \leq \eta_{\inf}(\alpha) + \varepsilon\right\} \geq \alpha$$

which implies that $(\xi\eta)_{\inf}(\alpha) \leq (\xi_{\inf}(\alpha) + \varepsilon)(\eta_{\inf}(\alpha) + \varepsilon)$. When $\xi_{\inf}(\alpha)\eta_{\inf}(\alpha) = 0$, letting $\varepsilon \to 0$, we obtain $(\xi\eta)_{\inf}(\alpha) = 0$. Otherwise, we have $\xi_{\inf}(\alpha) > 0$ and $\eta_{\inf}(\alpha) > 0$. Let $\varepsilon_0 \in (0, \varepsilon)$ be a small real number such that $\xi_{\inf}(\alpha) - \varepsilon_0 > 0$ and $\eta_{\inf}(\alpha) - \varepsilon_0 > 0$. Based on the relation

$$\left\{\xi\eta \leq \left(\xi_{\inf}(\alpha) - \varepsilon_0\right)\left(\eta_{\inf}(\alpha) - \varepsilon_0\right)\right\} \subseteq \left\{\xi \leq \xi_{\inf}(\alpha) - \varepsilon_0\right\} \cup \left\{\eta \leq \eta_{\inf}(\alpha) - \varepsilon_0\right\},$$

it follows from the independence and the monotonicity axiom that

$$\mathrm{Cr}\left\{\xi\eta \leq \left(\xi_{\inf}(\alpha) - \varepsilon_0\right)\left(\eta_{\inf}(\alpha) - \varepsilon_0\right)\right\}$$
$$\leq \mathrm{Cr}\left\{\xi \geq \xi_{\inf}(\alpha) - \varepsilon_0\right\} \vee \mathrm{Cr}\left\{\eta \leq \eta_{\inf}(\alpha) - \varepsilon_0\right\} < \alpha$$

which implies that $(\xi\eta)_{\inf}(\alpha) \geq (\xi_{\inf}(\alpha) - \varepsilon_0)(\eta_{\inf}(\alpha) - \varepsilon_0)$. In general, we get

$$\left(\xi_{\inf}(\alpha) - \varepsilon_0\right)\left(\eta_{\inf}(\alpha) - \varepsilon_0\right) \leq (\xi\eta)_{\inf}(\alpha) \leq \left(\xi_{\inf}(\alpha) + \varepsilon\right)\left(\eta_{\inf}(\alpha) + \varepsilon\right).$$

Letting $\varepsilon \to 0$, we obtain $(\xi\eta)_{\inf}(\alpha) = \xi_{\inf}(\alpha)\eta_{\inf}(\alpha)$.

(b) Since $\xi \leq 0, \eta \leq 0$, we have $-\xi \geq 0, -\eta \geq 0$. It follows from (a) that

$$(\xi\eta)_{\inf}(\alpha) = \left((-\xi)(-\eta)\right)_{\inf}(\alpha) = (-\xi)_{\inf}(\alpha)(-\eta)_{\inf}(\alpha)$$
$$= \xi_{\sup}(\alpha)\eta_{\sup}(\alpha).$$

(c) Since $\xi \geq 0$ and $-\eta \geq 0$, it follows from Theorems 4.5 and 4.12 that

$$(\xi\eta)_{\inf}(\alpha) = -\left(\xi(-\eta)\right)_{\sup}(\alpha) = -\xi_{\sup}(\alpha)(-\eta)_{\sup}(\alpha)$$
$$= \xi_{\sup}(\alpha)\eta_{\inf}(\alpha).$$

The proof is complete.   □

Example 4.16 Similarly, the independence condition cannot be removed in Theorem 4.14. Let us reconsider Example 4.7. It is easy to prove that

$$\xi_{\sup}(0.6) = 1, \qquad \xi_{\inf}(0.6) = 2, \qquad (-\xi)_{\sup}(0.6) = -2,$$
$$\eta_{\inf}(0.6) = 2, \qquad (-\eta)_{\sup}(0.6) = -2, \qquad (-\eta)_{\inf}(0.6) = -1,$$
$$(\xi\eta)_{\inf}(0.6) = 2, \qquad (\xi(-\eta))_{\inf}(0.6) = -2, \qquad ((-\xi)(-\eta))_{\inf}(0.6) = 2,$$

which implies that $(\xi\eta)_{inf}(0.6) \neq \xi_{inf}(0.6)\eta_{inf}(0.6)$, $(\xi(-\eta))_{inf}(0.6) \neq \xi_{sup}(0.6)$ $(-\eta)_{inf}(0.6)$, and $((-\xi)(-\eta))_{inf}(0.6) \neq (-\xi)_{sup}(0.6)(-\eta)_{sup}(0.6)$.

Theorem 4.15 (Li and Liu 2006b) *Suppose that ξ and η are independent fuzzy variables. Then for any $\alpha \in (0, 1]$, we have*

$$(\xi \wedge \eta)_{inf}(\alpha) = \xi_{inf}(\alpha) \wedge \eta_{inf}(\alpha). \qquad (4.12)$$

Proof According to Theorem 4.7, it is easy to prove that

$$(\xi \wedge \eta)_{inf}(\alpha) = -\big((-\xi) \vee (-\eta)\big)_{sup}(\alpha)$$
$$= -\big((-\xi)_{sup}(\alpha) \vee (-\eta)_{sup}(\alpha)\big)$$
$$= \xi_{inf}(\alpha) \wedge \eta_{inf}(\alpha).$$

The proof is complete.   □

Theorem 4.16 (Li and Liu 2006b) *Suppose that ξ and η are independent fuzzy variables. Then for any $\alpha \in (0, 1]$, we have*

$$(\xi \vee \eta)_{inf}(\alpha) = \xi_{inf}(\alpha) \vee \eta_{inf}(\alpha). \qquad (4.13)$$

Proof According to Theorem 4.6, it is easy to prove that

$$(\xi \vee \eta)_{inf}(\alpha) = -\big((-\xi) \wedge (-\eta)\big)_{sup}(\alpha)$$
$$= -\big((-\xi)_{sup}(\alpha) \wedge (-\eta)_{sup}(\alpha)\big)$$
$$= \xi_{inf}(\alpha) \vee \eta_{inf}(\alpha).$$

The proof is complete.   □

4.3   Chance-Constrained Programming Model

Generally speaking, the α-optimistic value denotes the maximum value the fuzzy variable can reach with credibility α. For fuzzy decision problems, since some decision-makers prefer to maximize the optimistic value of the objective subject to a set of chance constraints, Liu and Iwamura (1998a,b) proposed the following maximax chance-constrained programming model,

$$\begin{cases} \max & f(\boldsymbol{x}, \boldsymbol{\xi})_{sup}(\alpha) \\ \text{s.t.} & Cr\{g_i(\boldsymbol{x}, \boldsymbol{\xi}) \leq 0, \ i = 1, 2, \ldots, n\} \geq \beta \end{cases} \qquad (4.14)$$

where α and β are the predetermined confidence levels. On the other hand, the decision-makers can also maximize the pessimistic objective. In this case, Liu (1998) proposed the maximin chance-constrained programming model,

$$\begin{cases} \max & f(x, \boldsymbol{\xi})_{\inf}(\alpha) \\ \text{s.t.} & \text{Cr}\{g_i(x, \boldsymbol{\xi}) \le 0, \ i = 1, 2, \ldots, n\} \ge \beta. \end{cases} \tag{4.15}$$

For simplicity, we will call both the maximax model (4.14) and the maximin model (4.15) as the chance-constrained programming model when no ambiguity is possible.

Remark 4.7 Generally speaking, the confidence parameter β is selected to be greater than 0.5. In this case, it follows from Theorem 1.4 that

$$\text{Cr}\{g_i(x, \boldsymbol{\xi}) \le 0, \ i = 1, 2, \ldots, n\} \ge \beta$$

if and only if

$$\text{Cr}\{g_i(x, \boldsymbol{\xi}) \le 0\} \ge \beta, \quad i = 1, 2, \ldots, n.$$

Remark 4.8 The concepts of feasible solution, local optimal solution, and global optimal solution are given by Definitions 2.6, 2.7, and 2.8.

Remark 4.9 For the chance-constrained programming models, the feasibility and optimality depend on the value of confidence level β. It is possible that a solution is global optimal when $\beta = 0.8$, but even unfeasible when $\beta = 0.9$.

In what follows, we discuss the crisp equivalents for the maximax chance-constrained programming model (4.14) with the following conditions:

(c_1) confidence level $\beta > 0.5$;
(c_2) objective function $f(x, \boldsymbol{\xi}) = f_0(x) + f_1(x)\xi_1 + f_2(x)\xi_2 + \cdots + f_q(x)\xi_q$;
(c_3) constraint function $g_i(x, \boldsymbol{\xi}) = g_{i0}(x) + g_{i1}(x)\xi_1 + g_{i2}(x)\xi_2 + \cdots + g_{iq}(x)\xi_q$,

where f_k and g_{ik} are nonnegative functions for all $1 \le i \le n$ and $0 \le k \le q$.

Theorem 4.17 *Suppose that $\xi_i = (a_i, b_i, c_i)$, $i = 1, 2, \ldots, m$ are mutually independent triangular fuzzy variables. If model (4.14) satisfies conditions $(c_1$–$c_3)$, then it has the following crisp equivalent,*

$$\begin{cases} \max & f\left(x, \boldsymbol{\xi}_{\sup}(\alpha)\right) \\ \text{s.t.} & g_i\left(x, (3 - 2\beta)c - (2 - 2\beta)b\right) \le 0, \quad i = 1, 2, \ldots, n. \end{cases}$$

where $\boldsymbol{\xi}_{\sup}(\alpha) = ((\xi_1)_{\sup}(\alpha), (\xi_2)_{\sup}(\alpha), \ldots, (\xi_m)_{\sup}(\alpha))$.

Proof Since fuzzy variables $\xi_1, \xi_2, \ldots, \xi_m$ are mutually independent, it follows from the linearity of optimistic value that

$$f(x, \boldsymbol{\xi})_{\sup}(\alpha) = f\left(x, \boldsymbol{\xi}_{\sup}(\alpha)\right).$$

In addition, for each $i = 1, 2, \ldots, n$, it is easy to prove that $g_i(x, \boldsymbol{\xi})$ is a triangular fuzzy variable $(g_i(x, a), g_i(x, b), g_i(x, c))$. Since $\beta > 0.5$, we have

$$\text{Cr}\{g_i(\boldsymbol{x}, \boldsymbol{\xi}) \leq 0\} \geq \beta$$

if and only if $g_i(\boldsymbol{x}, (3 - 2\beta)\boldsymbol{c} - (2 - 2\beta)\boldsymbol{b}) \leq 0$. The proof is complete. □

Remark 4.10 In Theorem 4.17, if the triangular fuzzy variables are changed to be equipossible fuzzy variables, the crisp equivalent is

$$\begin{cases} \max & f\left(\boldsymbol{x}, \boldsymbol{\xi}_{\sup}(\alpha)\right) \\ \text{s.t.} & g_i(\boldsymbol{x}, \boldsymbol{b}) \leq 0, \quad i = 1, 2, \ldots, n. \end{cases}$$

If the triangular fuzzy variables are changed to be trapezoidal fuzzy variables, the crisp equivalent is

$$\begin{cases} \max & f\left(\boldsymbol{x}, \boldsymbol{\xi}_{\sup}(\alpha)\right) \\ \text{s.t.} & g_i\left(\boldsymbol{x}, (3 - 2\beta)\boldsymbol{d} - (2 - 2\beta)\boldsymbol{c}\right) \leq 0, \quad i = 1, 2, \ldots, n. \end{cases}$$

For normal fuzzy variables, the crisp equivalent is

$$\begin{cases} \max & f\left(\boldsymbol{x}, \boldsymbol{\xi}_{\sup}(\alpha)\right) \\ \text{s.t.} & g_i(\boldsymbol{x}, \boldsymbol{e}) \leq \lambda\left(g_i(\boldsymbol{x}, \sigma) - g_{i0}(\boldsymbol{x})\right), \quad i = 1, 2, \ldots, n \end{cases}$$

where $\lambda = (\sqrt{6}/\pi) \ln(1/\beta - 1)$.

Remark 4.11 The crisp equivalents for the maximin chance-constrained programming model can be proved similarly. In addition, if the objective function has the following form

$$f(\boldsymbol{x}, \boldsymbol{\xi}) = \max\{f_0(\boldsymbol{x}), f_1(\boldsymbol{x})\xi_1, f_2(\boldsymbol{x})\xi_2, \ldots, f_m(\boldsymbol{x})\xi_m\},$$
$$f(\boldsymbol{x}, \boldsymbol{\xi}) = \min\{f_0(\boldsymbol{x}), f_1(\boldsymbol{x})\xi_1, f_2(\boldsymbol{x})\xi_2, \ldots, f_m(\boldsymbol{x})\xi_m\},$$

these crisp equivalents still hold.

In many cases, there are multiple conflicting objectives. If the decision-maker prefers to get a solution with the maximum optimistic values for all objectives, we have the following multi-objective maximax chance-constrained programming model

$$\begin{cases} \max & \left[f_1(\boldsymbol{x}, \boldsymbol{\xi})_{\sup}(\alpha_1), f_2(\boldsymbol{x}, \boldsymbol{\xi})_{\sup}(\alpha_2), \ldots, f_p(\boldsymbol{x}, \boldsymbol{\xi})_{\sup}(\alpha_p)\right] \\ \text{s.t.} & \text{Cr}\{g_i(\boldsymbol{x}, \boldsymbol{\xi}) \leq 0, \ i = 1, 2, \ldots, n\} \geq \beta. \end{cases}$$

On the other hand, if the decision-maker prefers to get a solution with the maximum pessimistic values for all objectives, we have the following multi-objectives maximin chance-constrained programming model

$$\begin{cases} \max & \left[f_1(\boldsymbol{x}, \boldsymbol{\xi})_{\inf}(\alpha_1), f_2(\boldsymbol{x}, \boldsymbol{\xi})_{\inf}(\alpha_2), \ldots, f_p(\boldsymbol{x}, \boldsymbol{\xi})_{\inf}(\alpha_p)\right] \\ \text{s.t.} & \text{Cr}\{g_i(\boldsymbol{x}, \boldsymbol{\xi}) \leq 0, \ i = 1, 2, \ldots, n\} \geq \beta. \end{cases}$$

Remark 4.12 For the multi-objective chance-constrained programming models, we can also discuss the crisp equivalents when fuzzy parameters $\xi_1, \xi_2, \ldots, \xi_m$ are independent equipossible fuzzy variables, triangular fuzzy variables, trapezoidal fuzzy variables, and normal fuzzy variables.

4.4 Fuzzy Simulation

In order to solve the general chance-constrained programming models, this section introduces a fuzzy simulation technique for approximating the optimistic value and the pessimistic value

$$U_{\text{sup}} : x \rightarrow f(x, \xi)_{\text{sup}}(\alpha) \tag{4.16}$$

$$U_{\text{inf}} : x \rightarrow f(x, \xi)_{\text{inf}}(\alpha) \tag{4.17}$$

where f is a real function, and ξ is a fuzzy vector with joint credibility function v.

4.4.1 Optimistic Value Simulation

We first randomly generate vector y_i and calculate the credibility v_i for all $i = 1, 2, \ldots, N$. For any $r \in \Re$, according to the credibility inversion theorem, the credibility $\text{Cr}\{f(x, \xi) \geq r\}$ can be estimated by

$$C(r) = \begin{cases} \max\{v_i \mid f(x, y_i) \geq r\}, & \text{if } \max\{v_i \mid f(x, y_i) \geq r\} < 0.5 \\ \min\{1 - v_i \mid f(x, y_i) < r\}, & \text{if } \max\{v_i \mid f(x, y_i) \geq r\} \geq 0.5. \end{cases}$$

According to Definition 4.1, the α-optimistic value can be simulated as the maximal value of r satisfying $C(r) \geq \alpha$. Since C is a decreasing function, we can solve the maximal value by using a bisection search. The procedure is listed as follows.

Algorithm 4.1 (Fuzzy simulation for optimistic value)

Step 1. Initialize a small positive number ε.
Step 2. Randomly generate y_i with credibilities v_i for all $i = 1, 2, \ldots, N$.
Step 3. Calculate the minimum and maximum objective values

$$a = \min\{f(x, y_i) \mid 1 \leq i \leq N\}, \qquad b = \max\{f(x, y_i) \mid 1 \leq i \leq N\}.$$

Step 4. Set $r = (a + b)/2$.
Step 5. If $C(r) \geq \alpha$, set $a = r$. Otherwise, set $b = r$.
Step 6. If $b - a > \varepsilon$, go to step 4.
Step 7. Return $(a + b)/2$ as the α-optimistic value.

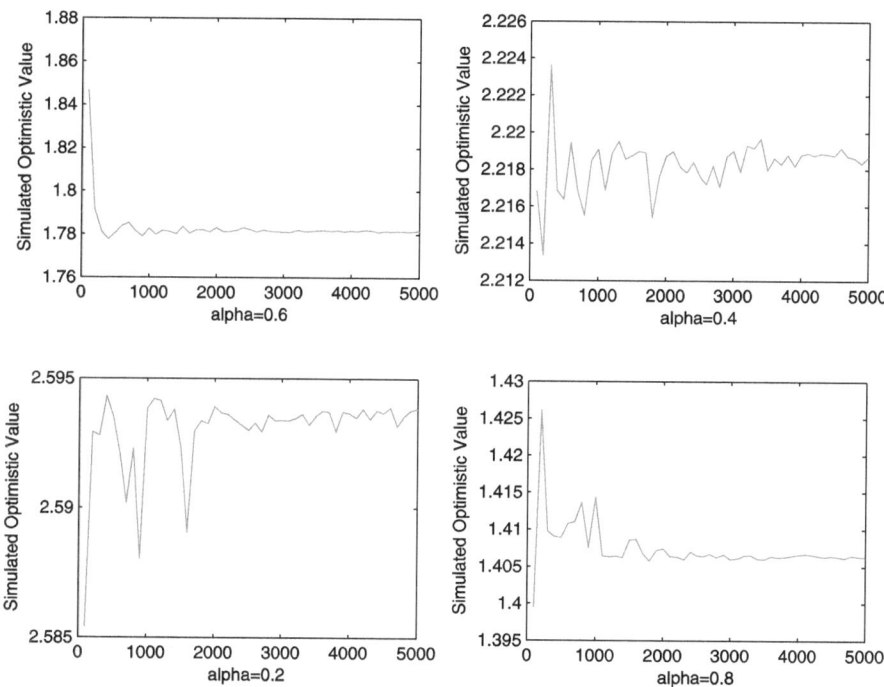

Fig. 4.15 Optimistic value simulation with variable parameter N

Example 4.17 For the triangular fuzzy variable $\xi = (1, 2, 3)$, we perform Algorithm 4.1 by changing N from 100 to 5000 with step of 100, and take α from $\{0.6, 0.4, 0.2, 0.8\}$. The simulated results are illustrated by Fig. 4.15. It is shown that when N is larger than 3500, the simulated results are stable.

Example 4.18 Taking $N = 3500$ and $\alpha = 0.4$, we perform Algorithm 4.1 on twenty fuzzy variables, including equipossible fuzzy variables, triangular fuzzy variables, trapezoidal fuzzy variables, normal fuzzy variables, and exponential fuzzy variables. We record the results by Table 4.1, and make comparisons with the exact values. The last column shows the relative error (3.25), which ranges from 0.00 % to 4.71 % and the average value is calculated to be 0.4 %. These results imply that the simulation algorithm can obtain a very satisfactory approximation for the optimistic value.

Example 4.19 Note that the optimistic value simulation is essentially a combination of the Monte Carlo simulation and the bisection search. Therefore, we may get different values if we perform the algorithm more than one times on the same fuzzy variable. In this example, we take $N = 3500$ and $\alpha = 0.4$, and perform Algorithm 4.1 fifty times on the triangular fuzzy variable $\xi = (-0.3, 1.8, 2.3)$. The results are recorded by Table 4.2. Compared with the exact value 1.9, the maximum relative error is 0.12 %.

Table 4.1 Simulation results on optimistic value

Fuzzy variables	Simulated value	Exact value	Relative error
(0.0, 1.0)	0.9988	1.0000	0.0012
(−1.0, 2.0)	1.9996	2.0000	0.0002
(3.5, 7.8)	7.7954	7.8000	0.0006
(−50, −10)	−10.0090	−10.0000	0.0009
(−0.3, 1.8, 2.3)	1.9000	1.9000	0.0000
(1.5, 3.0, 4.1)	3.2197	3.2200	0.0001
(10, 15, 20)	15.9969	16.0000	0.0002
(25, 40, 50)	41.9933	42.0000	0.0002
(1.0, 2.0, 3.0, 4.0)	3.1994	3.2000	0.0002
(3.1, 4.2, 4.5, 6.0)	4.7996	4.8000	0.0001
(2.4, 3.6, 3.7, 5.6)	4.0796	4.0800	0.0001
(10, 25, 30, 45)	32.9837	33.0000	0.0005
N(1.5, 1.0)	1.8150	1.8161	0.0001
N(2.5, 1.0)	2.8161	2.8161	0.0000
N(3.7, 2.1)	4.1583	4.3639	0.0471
N(6.0, 1.3)	6.3588	6.4110	0.0081
E(1.3)	0.4105	0.4110	0.0013
E(1.0)	0.3113	0.3161	0.0154
E(3.5)	1.1031	1.1065	0.0031
E(7.6)	2.3990	2.4027	0.0015

4.4.2 Pessimistic Value Simulation

The pessimistic value simulation has a similar process with the optimistic value simulation, which is also a combination of the Monte Carlo simulation and the bisection search. First, we randomly generate vectors y_i and calculate the credibilities v_i for all $i = 1, 2, \ldots, N$. Then for any $r \in \Re$, the credibility $\mathrm{Cr}\{f(x, \xi) \leq r\}$ can be estimated by

$$L(r) = \begin{cases} \max\{v_i \mid f(x, y_i) \leq r\}, & \text{if } \max\{v_i \mid f(x, y_i) \leq r\} < 0.5 \\ \min\{1 - v_i \mid f(x, y_i) > r\}, & \text{if } \max\{v_i \mid f(x, y_i) \leq r\} \geq 0.5. \end{cases}$$

According to Definition 4.2, the α-pessimistic value is the minimal value which satisfies $L(r) \geq \alpha$. Since $L(r)$ is an increasing function, we can calculate the minimal value by using the bisection search.

Algorithm 4.2 (Fuzzy simulation for pessimistic value)

Step 1. Initialize a small positive number ε.
Step 2. Randomly generate y_i with credibilities v_i for all $i = 1, 2, \ldots, N$.

Table 4.2 Simulation on the 0.4-optimistic value of $\xi = (-0.3.1.8, 2.3)$

	Simulated value	Error		Simulated value	Error
1	1.8974	0.0014	26	1.8997	0.0002
2	1.8993	0.0004	27	1.8998	0.0001
3	1.8998	0.0001	28	1.8985	0.0008
4	1.8993	0.0004	29	1.8990	0.0005
5	1.9002	0.0001	30	1.8986	0.0008
6	1.8982	0.0010	31	1.8982	0.0009
7	1.8999	0.0001	32	1.8991	0.0005
8	1.8984	0.0008	33	1.9000	0.0000
9	1.8978	0.0012	34	1.8991	0.0005
10	1.9000	0.0000	35	1.8996	0.0002
11	1.8990	0.0005	36	1.8984	0.0009
12	1.9001	0.0001	37	1.8991	0.0005
13	1.8963	0.0019	38	1.8979	0.0011
14	1.9001	0.0001	39	1.8997	0.0001
15	1.8988	0.0006	40	1.8995	0.0003
16	1.8991	0.0005	41	1.8988	0.0007
17	1.8996	0.0002	42	1.8993	0.0004
18	1.8999	0.0001	43	1.8994	0.0003
19	1.8995	0.0003	44	1.8998	0.0001
20	1.8993	0.0004	45	1.8997	0.0002
21	1.8987	0.0007	46	1.8977	0.0012
22	1.8997	0.0001	47	1.9001	0.0001
23	1.9000	0.0000	48	1.8985	0.0008
24	1.8991	0.0005	49	1.8997	0.0002
25	1.8996	0.0002	50	1.8986	0.0007

Step 3. Calculate the minimum and maximum objective values

$$a = \min\{ f(x, y_i) \mid 1 \leq i \leq N\}, \qquad b = \max\{ f(x, y_i) \mid 1 \leq i \leq N\}.$$

Step 4. Set $r = (a + b)/2$.
Step 5. If $L(r) \geq \alpha$, set $b = r$. Otherwise, set $a = r$.
Step 6. If $b - a > \varepsilon$, go to step 4.
Step 7. Return $(a + b)/2$ as the α-pessimistic value.

Example 4.20 For the triangular fuzzy variable $\xi = (1, 2, 3)$, we perform Algorithm 4.2 by changing N from 100 to 5000 with step of 100, and take α in the set of $\{0.6, 0.4, 0.2, 0.8\}$. The simulated results are illustrated by Fig. 4.16. It is shown that when $N \geq 2500$, the simulated results are stable.

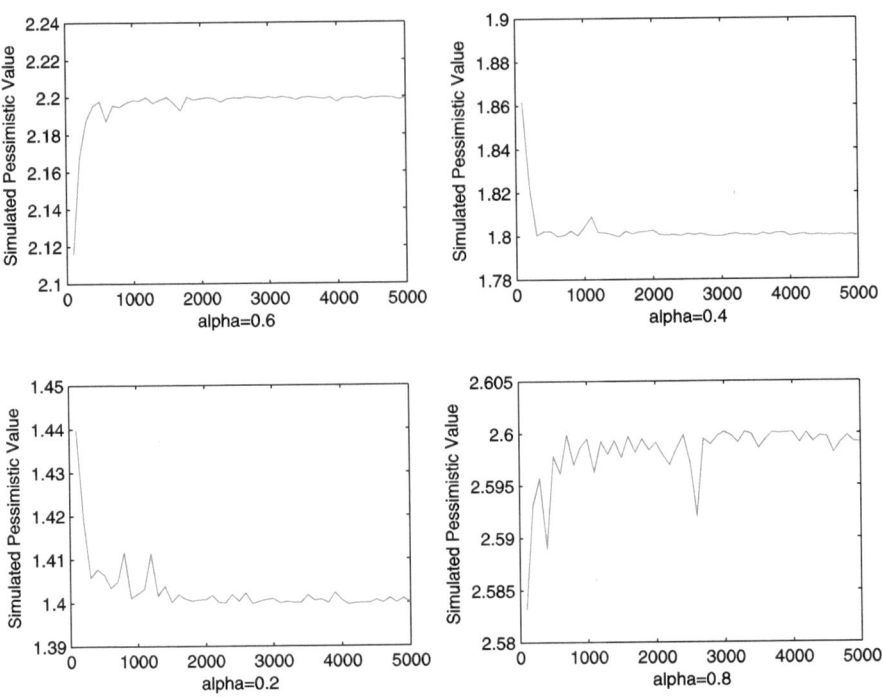

Fig. 4.16 The pessimistic value simulation with variable parameter N

Example 4.21 Taking $N = 2500$ and $\alpha = 0.6$, we perform Algorithm 4.2 on twenty fuzzy variables, including equipossible fuzzy variables, triangular fuzzy variables, trapezoidal fuzzy variables, normal fuzzy variables, and exponential fuzzy variables. We record the results by Table 4.3, and make comparisons with the exact values. The relative error ranges from 0.00 % to 4.72 % and the average value is 0.45 %, which implies that the fuzzy simulation can obtain a very satisfactory approximation for the pessimistic value.

Example 4.22 In this example, we take $N = 2500$ and $\alpha = 0.6$, and perform Algorithm 4.2 fifty times on fuzzy variable $\xi = (-0.3, 1.8, 2.3)$. The results are recorded by Table 4.4. Compared with the exact value 1.9, the maximum relative error is 0.21 %.

4.5 Applications

This section applies the maximax chance-constrained programming model to the fuzzy portfolio selection problem. See Examples 2.4 and 2.5. If the decision-maker prefers a portfolio with larger optimistic return under certain chance constraints, we get the following maximax chance-constrained programming model,

Table 4.3 Simulation results on pessimistic value

Fuzzy variables	Simulated value	Exact value	Relative error
(0.0, 1.0)	0.9988	1.0000	0.0012
(−1.0, 2.0)	1.9985	2.0000	0.0008
(3.5, 7.8)	7.7997	7.8000	0.0001
(−50, −10)	−10.0050	−10.0000	0.0005
(−0.3, 1.8, 2.3)	1.8998	1.9000	0.0001
(1.5, 3.0, 4.1)	3.2189	3.2200	0.0003
(10, 15, 20)	15.9635	16.0000	0.0023
(25, 40, 50)	41.9735	42.0000	0.0006
(1.0, 2.0, 3.0, 4.0)	3.1966	3.2000	0.0011
(3.1, 4.2, 4.5, 6.0)	4.7925	4.8000	0.0016
(2.4, 3.6, 3.7, 5.6)	4.0778	4.0800	0.0005
(10, 25, 30, 45)	32.9999	33.0000	0.0000
$N(1.5, 1.0)$	1.8149	1.8161	0.0007
$N(2.5, 1.0)$	2.8164	2.8161	0.0001
$N(3.7, 2.1)$	4.1581	4.3639	0.0472
$N(6.0, 1.3)$	6.3483	6.4110	0.0098
$E(1.3)$	0.4076	0.4110	0.0081
$E(1.0)$	0.3126	0.3161	0.0111
$E(3.5)$	1.1018	1.1065	0.0042
$E(7.6)$	2.4011	2.4027	0.0006

$$\begin{cases} \max & (\xi_1 x_1 + \xi_2 x_2 + \cdots + \xi_m x_m)_{\sup}(\alpha) \\ \text{s.t.} & \mathrm{Cr}\{\xi_{i_1} x_{i_1} + \xi_{i_2} x_{i_2} + \cdots + \xi_{i_s} x_{i_s} \geq \gamma\} \geq \beta \\ & x_1 + x_2 + \cdots + x_m = 1 \\ & x_i \geq 0, \quad i = 1, 2, \ldots, m \end{cases} \qquad (4.18)$$

where the first constraint denotes that the return from stocks $i_1, i_2, \ldots, i_s$ should be larger than γ with credibility β.

Example 4.23 Suppose that there are four stocks with independent normal fuzzy returns (see Table 3.3). Taking $\alpha = 0.8$, $\beta = 0.5$, $\gamma = 1.2$ and $i_1 = 3$, $i_2 = 4$ in model (4.18), we get the following linear programming model

$$\begin{cases} \max & 0.0191x_1 + 0.111x_2 + 0.0948x_3 - 0.5133x_4 \\ \text{s.t.} & 1.5x_3 + x_4 \geq 1.2 \\ & x_1 + x_2 + x_3 + x_4 = 1 \\ & x_1, x_2, x_3, x_4 \geq 0. \end{cases}$$

Table 4.4 Simulation on the 0.6-pessimistic value of $\xi = (-0.3.1.8, 2.3)$

	Simulated value	Error		Simulated value	Error
1	1.9001	0.0000	26	1.8992	0.0004
2	1.8986	0.0007	27	1.8988	0.0006
3	1.8997	0.0002	28	1.8982	0.0009
4	1.8971	0.0015	29	1.9000	0.0000
5	1.8997	0.0001	30	1.8997	0.0002
6	1.8988	0.0007	31	1.8990	0.0006
7	1.8999	0.0001	32	1.8993	0.0004
8	1.8983	0.0009	33	1.8960	0.0021
9	1.8998	0.0001	34	1.8971	0.0015
10	1.8985	0.0008	35	1.8984	0.0008
11	1.8975	0.0013	36	1.8969	0.0016
12	1.8974	0.0014	37	1.8989	0.0006
13	1.8990	0.0005	38	1.8976	0.0013
14	1.8988	0.0006	39	1.9000	0.0000
15	1.8992	0.0004	40	1.8989	0.0006
16	1.8994	0.0003	41	1.8988	0.0007
17	1.8998	0.0001	42	1.9000	0.0000
18	1.8997	0.0001	43	1.8985	0.0008
19	1.8995	0.0003	44	1.8948	0.0027
20	1.9000	0.0000	45	1.8974	0.0014
21	1.8999	0.0001	46	1.8994	0.0003
22	1.8992	0.0004	47	1.8992	0.0004
23	1.8995	0.0003	48	1.8981	0.0010
24	1.8977	0.0012	49	1.8989	0.0006
25	1.8993	0.0004	50	1.8997	0.0002

We use the Matlab function *Linprog* to solve the linear programming model. The optimal objective value is 0.0980, and the optimal portfolio is

$$x_1 = 0, \qquad x_2 = 0.2, \qquad x_3 = 0.8, \qquad x_4 = 0.$$

Note that the optimal portfolio distributes no capital on the first and the fourth stocks.

Example 4.24 In this example, the chance-constrained programming model (4.18) is applied to the data shown in Table 3.4, which is composed of two triangular fuzzy variables and two normal fuzzy variables. Taking $\alpha = 0.8$, $\beta = 0.7$, $\gamma = 0.6$ and $i_1 = 3$, $i_2 = 4$, we get the following model

$$\begin{cases} \max & (\xi_1 x_1 + \xi_2 x_2 + \xi_3 x_3 + \xi_4 x_4)_{\sup}(\alpha) \\ \text{s.t.} & Cr\{\xi_3 x_3 + \xi_4 x_4 \ge 0.6\} \ge 0.7 \\ & x_1 + x_2 + x_3 + x_4 = 1 \\ & x_1, x_2, x_3, x_4 \ge 0. \end{cases}$$

Since the model has no crisp equivalent, the genetic algorithm is used to solve the model, which is coded in Matlab programming language under the running environment: a Windows 7 platform of personal computer with processor speed 2.4 GHz and memory size 2 GB.

Take $N = 3000$, $G = 30$, $P_c = 0.4$, $P_m = 0.2$ and $pop\text{-}size = 100$. A run of the genetic algorithm shows that the optimal portfolio is

$$x_1 = 0.1393, \qquad x_2 = 0.0885, \qquad x_3 = 0.4075, \qquad x_4 = 0.3647,$$

and the maximum optimistic value is 1.6622.

References

Charnes A, Cooper WW (1961) Management models and industrial applications of linear programming. Wiley, New York

Ke H, Ma WM, Gao X, Xu WH (2010) New fuzzy models for time-cost trade-off problem. Fuzzy Optim Decis Mak 9(2):219–231

Li X, Liu B (2006b) The independence of fuzzy variables with applications. Int J Nat Sci Technol 1(1):95–100

Li X, Qin ZF, Yang L (2010b) A chance-constrained portfolio selection model with risk constraints. Appl Math Comput 217:949–951

Liu B (1998) Minimax chance constrained programming models for fuzzy decision systems. Inf Sci 112(1–4):25–38

Liu B (2004) Uncertainty theory: an introduction to its axiomatic foundations. Springer, Berlin

Liu B, Iwamura K (1998a) Chance constrained programming with fuzzy parameters. Fuzzy Sets Syst 94(2):227–237

Liu B, Iwamura K (1998b) A note on chance constrained programming with fuzzy coefficients. Fuzzy Sets Syst 100(1–3):229–233

Liu LZ, Li YZ (2006) The fuzzy quadratic assignment problem with penalty: new models and genetic algorithm. Appl Math Comput 174(2):1229–1244

Shao Z, Ji XY (2006) Fuzzy multi-product constraint newsboy problem. Appl Math Comput 180(1):7–15

Zheng Y, Liu B (2006) Fuzzy vehicle routing model with credibility measure and its hybrid intelligent algorithm. Appl Math Comput 176(2):673–683

Zhou J, Liu B (2007) Modeling capacitated location-allocation problem with fuzzy demands. Comput Ind Eng 53(3):454–468

Chapter 5
Entropy Optimization Model

Fuzzy entropy is used to characterize the uncertainty on the possible values of fuzzy variables, which has been studied by many researchers such as Bhandari and Pal (1993), De Luca and Termini (1972), Kaufmann (1975), Kosko (1986), Liu (1992), Pal and Pal (1992), Pal and Bezdek (1994), Szmidt and Kacprzyk (2001), and Yager (1979). Within the framework of credibility theory, Li and Liu (2008a) presented a Shannon-like entropy for both discrete fuzzy variable and continuous fuzzy variable. Li and Liu (2007) proposed the maximum entropy principle, and proved that out of all the credibility functions with fixed expected value and variance, the normal credibility function has the maximum entropy. Based on the concept of fuzzy entropy, Li et al. (2011) proposed an entropy optimization model by minimizing the uncertainty of the fuzzy objective under certain expected constraints.

This chapter mainly includes the definition of fuzzy entropy, maximum entropy theorems, entropy optimization model and its crisp equivalents, fuzzy simulation, and applications in portfolio selection problem.

5.1 Entropy

This section introduces the definitions of entropy for discrete fuzzy variable and continuous fuzzy variable, respectively.

Definition 5.1 (Li and Liu 2008a) Suppose that ξ is a discrete fuzzy variable taking values in $\{x_1, x_2, \ldots\}$. Then its entropy is defined by

$$H[\xi] = \sum_{i=1}^{\infty} S\big(\mathrm{Cr}\{\xi = x_i\}\big) \tag{5.1}$$

where $S(t) = -t \ln t - (1 - t) \ln(1 - t)$.

It is easy to prove that the continuous and differentiable function $S(t)$ has the following properties (see Fig. 5.1):

X. Li, *Credibilistic Programming*, Uncertainty and Operations Research,
DOI 10.1007/978-3-642-36376-4_5, © Springer-Verlag Berlin Heidelberg 2013

Fig. 5.1 The shape of
function $S(t) = -t \ln t - (1-t)\ln(1-t)$

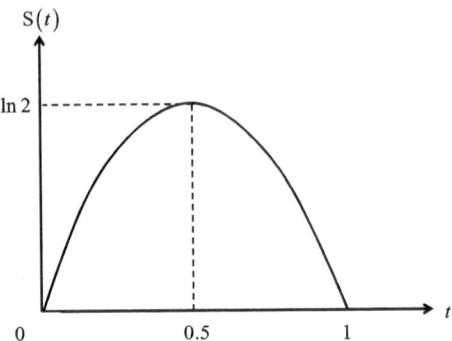

(a) it is a nonnegative, unimodal and concave function;
(b) it takes the minimum value zero at $t = 0$ and $t = 1$;
(c) it takes the maximum value $\ln 2$ at $t = 0.5$.

Remark 5.1 It is clear that the fuzzy entropy does not depend on the actual values
that the fuzzy variable takes, but only depends on the credibilities.

Example 5.1 Take $(\Theta, \mathcal{A}, \mathrm{Cr})$ to be a credibility space $\{\theta_1, \theta_2, \theta_3\}$ with $\mathrm{Cr}\{\theta_1\} = 0.4$, $\mathrm{Cr}\{\theta_2\} = 0.6$, and $\mathrm{Cr}\{\theta_3\} = 0.2$. Define two fuzzy variables

$$\xi = \begin{cases} 1, & \text{if } \theta = \theta_1 \\ 0, & \text{if } \theta = \theta_2 \\ -1, & \text{if } \theta = \theta_3, \end{cases} \qquad \eta = \begin{cases} 5, & \text{if } \theta = \theta_1 \\ 3, & \text{if } \theta = \theta_2 \\ -2, & \text{if } \theta = \theta_3. \end{cases}$$

It is clear that $\xi \neq \eta$ since they take different values on each point θ. However,
according to Definition 5.1, we have

$$S(0.4) = -0.4\ln 0.4 - 0.6\ln 0.6 = 0.6730,$$

$$S(0.6) = -0.6\ln 0.6 - 0.4\ln 0.4 = 0.6730,$$

$$S(0.2) = -0.2\ln 0.2 - 0.8\ln 0.8 = 0.3570,$$

which implies that the entropy is

$$H[\xi] = H[\eta] = 0.6730 + 0.6730 + 0.3570 = 1.7030.$$

This example tells us that fuzzy variables may have the same entropy even if they
have different credibility functions.

Example 5.2 Let ξ be a discrete fuzzy variable with credibility function

$$v(x_i) = \begin{cases} 1/2, & \text{if } i = 1 \\ 1/2(i-1), & \text{if } i \geq 2. \end{cases}$$

It is easy to prove that $S(v(x_1)) = S(v(x_2)) = \ln 2$, and for all $i \geq 3$, we have

$$S\big(v(x_i)\big) = -\frac{1}{2(i-1)}\ln\frac{1}{2(i-1)} - \left(1 - \frac{1}{2(i-1)}\right)\ln\left(1 - \frac{1}{2(i-1)}\right)$$

$$\geq \frac{1}{2(i-1)}\ln\big(2(i-1)\big)$$

$$\geq \frac{1}{2(i-1)}.$$

According to Definition 5.1, the entropy of fuzzy variable ξ is

$$H[\xi] = \sum_{i=1}^{\infty} S\big(v(x_i)\big) = +\infty.$$

Theorem 5.1 (Li and Liu 2008a) *Suppose that ξ is a discrete fuzzy variable taking values in $\{x_1, x_2, \ldots\}$. Then we have*

$$H[\xi] \geq 0$$

and the equality holds if and only if ξ is a constant number.

Proof It follows from the nonnegativity of function $S(t)$ that $H[\xi] \geq 0$. Furthermore, the equality holds if and only if $v(x_i) = 0$ or $v(x_i) = 1$ for all $i = 1, 2, \ldots$. According to the credibility extension condition, there exists one and only one index i with $v(x_i) = 1$ and $v(x_j) = 0$ for all $j \neq i$. That is, ξ degenerates to the constant number x_i. The proof is complete. □

This theorem states that the entropy of a discrete fuzzy variable reaches its minimum value zero when it degenerates to a constant number. In this case, there is no uncertainty on it.

Theorem 5.2 (Li and Liu 2008a) *Suppose that ξ is a simple fuzzy variable taking values in $\{x_1, x_2, \ldots, x_n\}$. Then we have*

$$H[\xi] \leq n \ln 2$$

and the equality holds if and only if $v(x_i) = 0.5$ for all $i = 1, 2, \ldots, n$.

Proof Since function $S(t)$ reaches its maximum value $\ln 2$ at $t = 0.5$, it is easy to prove that

$$H[\xi] = \sum_{i=1}^{n} S\big(v(x_i)\big) \leq n \ln 2$$

and the equality holds if and only if $v(x_i) = 0.5$ for all $i = 1, 2, \ldots, n$. The proof is complete. □

This theorem states that the entropy of a simple fuzzy variable reaches its maximum value when it has an equipossible credibility function. In this case, there is no preference among all values that the fuzzy variable will take. That is, the uncertainty has the maximum value.

Definition 5.2 (Li and Liu 2008a) Suppose that ξ is a continuous fuzzy variable with credibility function ν. Then its entropy is defined by

$$H[\xi] = \int_{-\infty}^{+\infty} S\big(\nu(x)\big)\,dx \qquad (5.2)$$

where $S(t) = -t\ln t - (1-t)\ln(1-t)$.

Example 5.3 Let ξ be an equipossible fuzzy variable taking values in $[a, b]$. Denote ν as its credibility function. Then for any $x \in [a, b]$, we have

$$S\big(\nu(x)\big) = -0.5\ln 0.5 - (1 - 0.5)\ln(1 - 0.5) = \ln 2,$$

which implies that the entropy is

$$H[\xi] = \int_a^b S\big(\nu(x)\big)\,dx = (b - a)\ln 2.$$

Example 5.4 Let ξ be a triangular fuzzy variable (a, b, c) with credibility function ν. According to the theorem of integration by parts, we have

$$\int_a^b S\big(\nu(x)\big)\,dx = \int_a^b S\big((x - a)/2(b - a)\big)\,dx = (b - a)/2,$$

$$\int_b^c S\big(\nu(x)\big)\,dx = \int_b^c S\big((c - x)/2(c - b)\big)\,dx = (c - b)/2,$$

which implies that the entropy is

$$H[\xi] = \int_a^b S\big(\nu(x)\big)\,dx + \int_b^c S\big(\nu(x)\big)\,dx = (c - a)/2.$$

Example 5.5 Let ξ be a trapezoidal fuzzy variable (a, b, c, d). According to the theorem of integration by parts, we have

$$\int_a^b S\big(\nu(x)\big)\,dx = \int_a^b S\big((x - a)/2(b - a)\big)\,dx = (b - a)/2,$$

$$\int_b^c S\big(\nu(x)\big)\,dx = \int_b^c S(0.5)\,dx = (c - b)\ln 2,$$

$$\int_c^d S\big(\nu(x)\big)\,dx = \int_c^d S\big((d - x)/2(d - c)\big)\,dx = (d - c)/2,$$

which implies that the entropy is

$$H[\xi] = \int_a^b S(v(x)) \, dx + \int_b^c S(v(x)) \, dx + \int_c^d S(v(x)) \, dx$$
$$= (b-a)/2 + (c-b)\ln 2 + (d-c)/2.$$

Example 5.6 Let $\xi = EXP(m)$ be an exponential fuzzy variable. For any $x \geq 0$, it is easy to prove that

$$S(v(x)) = \ln(1 + \exp(t)) - t\exp(t)/(1 + \exp(t))$$

where $t = \pi x / \sqrt{6}m$. According to Definition 5.2, we have

$$H[\xi] = \frac{\sqrt{6}m}{\pi} \int_0^{+\infty} \ln(1 + \exp(t)) - t\exp(t)/(1 + \exp(t)) \, dt = \frac{\pi m}{\sqrt{6}}.$$

This example tells us that the entropy for an exponential fuzzy variable is proportional to parameter m.

Example 5.7 Let $\xi = N(e, \sigma)$ be a normal fuzzy variable. For any $x \geq 0$, it is easy to prove that

$$S(v(x)) = \ln(1 + \exp(t)) - t\exp(t)/(1 + \exp(t))$$

where $t = \pi|x - e|/\sqrt{6}\sigma$. According to Definition 5.2, we have

$$H[\xi] = \frac{2\sqrt{6}\sigma}{\pi} \int_0^{+\infty} \ln(1 + \exp(t)) - t\exp(t)/(1 + \exp(t)) \, dt = \frac{\sqrt{6}\pi\sigma}{3}.$$

Therefore, the entropy for a normal fuzzy variable is proportional to the standard variance, but has no relation with its expected value.

Theorem 5.3 (Li and Liu 2008a) *Let ξ be a continuous fuzzy variable. Then we have*

$$H[\xi] > 0.$$

Proof For any $0 < \varepsilon < 0.5$, according to the credibility extension condition, there is a real number c with $v(c) > \varepsilon$. It follows from the continuity that there is a number $\delta > 0$ such that $v(x) \geq \varepsilon$ for all $|x - c| \leq \delta$. Then we have

$$H[\xi] \geq \delta \times S(\varepsilon) > 0.$$

The proof is complete. $\qquad\square$

Theorem 5.4 (Li and Liu 2008a) *Let ξ be a continuous fuzzy variable taking values in the interval $[a, b]$. Then we have*

$$H[\xi] \leq (b - a) \ln 2 \tag{5.3}$$

and the equality holds if and only if ξ is an equipossible fuzzy variable.

Proof Since function $S(t)$ reaches its maximum $\ln 2$ at $t = 0.5$, we have

$$H[\xi] \leq \int_a^b \ln 2 \, dx = (b - a) \ln 2,$$

and the equality holds if and only if $v(x) = 0.5$ for all $x \in [a, b]$, that is, ξ is an equipossible fuzzy variable. The proof is complete. □

Theorem 5.5 (Li and Liu 2008a) *Let ξ be a continuous fuzzy variable. Then for any real numbers a and b, we have*

$$H[a\xi + b] = |a| H[\xi]. \tag{5.4}$$

Proof Suppose that fuzzy variables ξ and $a\xi + b$ have credibility functions v and μ, respectively. If $a = 0$, then (5.4) is trivial. Otherwise, for any $x \in \Re$, it is easy to prove that

$$\mu(x) = \mathrm{Cr}\{a\xi + b = x\} = \mathrm{Cr}\{\xi = (x - b)/a\} = v\big((x - b)/a\big).$$

It follows from the definition of entropy that

$$H[a\xi + b] = \int_{-\infty}^{+\infty} S\big(v(x - b)/a\big) \, dx$$

$$= \int_{-\infty}^{+\infty} |a| S\big(v(x)\big) \, dx$$

$$= |a| H[\xi].$$

The theorem is proved. □

5.2 Maximum Entropy Principle

Let ξ be a fuzzy variable with some given information, for example, the values of expected value and variance. Generally speaking, there is an infinite number of credibility functions satisfying this information, and each of these has a different amount of entropy associated with it. The problem is which one we should take? In order to answer this problem, Li and Liu (2007) proposed the *maximum entropy principle: out of all the credibility functions satisfying the given constraints, choose the one that has the maximum entropy.*

The reason for this principle is quite simple. If we choose a credibility function with a smaller entropy, we may have used some additional information consciously or unconsciously. However, it is not correct to use such information because this is not given to us.

Theorem 5.6 (Li and Liu 2007) *Let ξ be a continuous nonnegative fuzzy variable with second moment m^2. Then we have*

$$H[\xi] \le \pi m/\sqrt{6} \qquad (5.5)$$

and the maximum entropy attains if ξ is an exponential fuzzy variable.

Proof Assume that continuous function ν is the credibility function of ξ. The proof is based on the following two steps.

Step 1. Suppose that ν is a decreasing function on $[0, +\infty)$. In this case, according to the credibility inversion theorem, we have $\nu(0) = 0.5$ and $\mathrm{Cr}\{\xi \ge x\} = \nu(x)$ for any $x > 0$. Thus the second moment is

$$E[\xi^2] = \int_0^{+\infty} \mathrm{Cr}\{\xi^2 \ge x\}\mathrm{d}x$$

$$= \int_0^{+\infty} 2x\mathrm{Cr}\{\xi \ge x\}\mathrm{d}x$$

$$= \int_0^{+\infty} 2x\nu(x)\mathrm{d}x.$$

The maximum entropy credibility function ν should maximize the entropy

$$-\int_0^{+\infty} \left(\nu(x)\ln\nu(x) + \left(1 - \nu(x)\right)\ln\left(1 - \nu(x)\right)\right)\mathrm{d}x$$

subject to the moment constraint

$$\int_0^{+\infty} 2x\nu(x)\mathrm{d}x = m^2.$$

The Lagrangian is

$$L \equiv -\int_0^{+\infty} \left(\nu(x)\ln\nu(x) + \left(1 - \nu(x)\right)\ln\left(1 - \nu(x)\right)\right)\mathrm{d}x$$
$$- \lambda\left(\int_0^{+\infty} 2x\nu(x)\mathrm{d}x - m^2\right).$$

Euler-Lagrange equation tells us that the maximum entropy credibility function meets $\ln \nu(x) - \ln(1 - \nu(x)) + 2\lambda x = 0$, and has the form

$$\nu(x) = 1/(1 + \exp(2\lambda x)).$$

Substituting it into the moment constraint, we get $\lambda = \pi/(2\sqrt{6}m)$ and then

$$\nu(x) = 1/(1 + \exp(\pi x/(\sqrt{6}m))), \quad x \geq 0,$$

which is just the exponential credibility function with second moment m^2, and the entropy is $\pi m/\sqrt{6}$.

Step 2. Let ξ be a general fuzzy variable with second moment m^2. Now we define a fuzzy variable η via credibility function

$$\mu(x) = \sup_{y \geq x} \nu(y), \quad x \geq 0.$$

Then $\mu(x)$ is a decreasing function on $[0, +\infty)$, and

$$\text{Cr}\{\eta^2 \geq x\} = \sup_{y \geq \sqrt{x}} \mu(y) = \sup_{y \geq \sqrt{x}} \sup_{z \geq y} \nu(z) = \sup_{y \geq \sqrt{x}} \nu(y) \leq \text{Cr}\{\xi^2 \geq x\}$$

for any $x > 0$. Thus we have

$$E[\eta^2] = \int_0^{+\infty} \text{Cr}\{\eta^2 \geq x\}\mathrm{d}x \leq \int_0^{+\infty} \text{Cr}\{\xi^2 \geq x\}\mathrm{d}x = m^2.$$

It follows from $\nu(x) \leq \mu(x)$ and step 1 that

$$H[\xi] \leq H[\eta] \leq \pi\sqrt{E[\eta^2]/6} \leq \pi m/\sqrt{6}.$$

The proof is complete.   □

Theorem 5.7 (Li and Liu 2007) *Let ξ be a continuous fuzzy variable with expected value e and variance σ^2. Then we have*

$$H[\xi] \leq \sqrt{6}\pi\sigma/3 \tag{5.6}$$

and the maximum entropy attains if ξ is a normal fuzzy variable.

Proof Assume that continuous function ν is the credibility function of ξ. The proof is based on the following two steps.

Step 1. Suppose that $\nu(x)$ is a unimodal and symmetric function about $x = e$. For any $x \geq 0$, it follows from Theorem 1.3 that

$$\text{Cr}\{(\xi - e)^2 \geq x\} = \text{Cr}\{(\xi - e \geq \sqrt{x}) \cup (\xi - e \leq -\sqrt{x})\}$$
$$= \text{Cr}\{\xi - e \geq \sqrt{x}\}.$$

Then, according to the definition of variance, we have

$$V[\xi] = \int_0^{+\infty} \mathrm{Cr}\{\xi - e \geq \sqrt{x}\}dx$$

$$= \int_e^{+\infty} 2(x - e)\mathrm{Cr}\{\xi \geq x\}dx$$

$$= \int_e^{+\infty} 2(x - e)\nu(x)dx.$$

The maximum entropy credibility function ν should maximize the entropy

$$H[\xi] = -\int_e^{+\infty} 2\big(\nu(x)\ln\nu(x) + (1 - \nu(x))\ln(1 - \nu(x))\big)dx$$

subject to the variance constraint. The Lagrangian is

$$L \equiv -\int_e^{+\infty} 2\big(\nu(x)\ln\nu(x) + (1 - \nu(x))\ln(1 - \nu(x))\big)dx$$

$$- \lambda\left(\int_e^{+\infty} 2(x - e)\nu(x)dx - \sigma^2\right).$$

Euler-Lagrange equation gives that the maximum entropy credibility function meets the following equation

$$\ln\nu(x) - \ln(1 - \nu(x)) + \lambda(x - e) = 0,$$

and has the form $\nu(x) = 1/(1 + \exp(\lambda(x - e)))$. Substituting it into the variance constraint, we get

$$\nu(x) = 1/\big(1 + \exp(\pi|x - e|/(\sqrt{6}\sigma))\big), \quad x \in \Re$$

which is just the normal credibility function with expected value e and variance σ^2, and the entropy is $\sqrt{6}\pi\sigma/3$.

Step 2. Let ξ be a general fuzzy variable with expected value e and variance σ^2. We define a fuzzy variable η by the credibility function

$$\mu(x) = \begin{cases} \sup_{y \leq x}\big(\nu(y) \vee \nu(2e - y)\big), & \text{if } x \leq e \\ \sup_{y \geq x}\big(\nu(y) \vee \nu(2e - y)\big), & \text{if } x > e. \end{cases} \tag{5.7}$$

It is easy to prove that $\mu(x)$ is a unimodal and symmetric function about $x = e$. Furthermore, for any $r > 0$, we have

$$\mathrm{Cr}\{(\eta - e)^2 \geq r\} = \sup_{x \geq e + \sqrt{r}} \mu(x) = \sup_{x \geq e + \sqrt{r}} \sup_{y \geq x}\big(\nu(y) \vee \nu(2e - y)\big)$$

$$= \sup_{x \geq e + \sqrt{r}} \left(v(x) \vee v(2e - x) \right) = \sup_{(x-e)^2 \geq r} v(x)$$

$$\leq \mathrm{Cr}\{ (\xi - e)^2 \geq r \}$$

which implies that

$$V[\eta] = \int_0^{+\infty} \mathrm{Cr}\{ (\eta - e)^2 \geq r \} \mathrm{d}r \leq \int_0^{+\infty} \mathrm{Cr}\{ (\xi - e)^2 \geq r \} \mathrm{d}r = \sigma^2.$$

It follows from $v(x) \leq \mu(x)$ and step 1 that

$$H[\xi] \leq H[\eta] \leq \pi \sqrt{6V[\eta]}/3 \leq \sqrt{6}\pi\sigma/3.$$

The proof is complete.   □

5.3 Entropy Optimization Model

For fuzzy decision problems, some decision-makers prefer the solution which has the minimum uncertainty on possible values of the objective function. If we use entropy to denote the uncertainty, we get the following entropy optimization model with expected constraints,

$$\begin{cases} \min & H[f(x, \xi)] \\ \text{s.t.} & E[f(x, \xi)] \geq \alpha \\ & E[g_i(x, \xi)] \leq 0, \quad i = 1, 2, \ldots, n \end{cases} \tag{5.8}$$

where α is a predetermined constant.

Remark 5.2 The concepts of feasible solution, local optimal solution, and global optimal solution are given by Definitions 2.6, 2.7, and 2.8.

In what follows, we discuss the crisp equivalents for the entropy optimization model satisfying the following conditions:

(c_1) objective function $f(x, \xi) = f_0(x) + f_1(x)\xi_1 + f_2(x)\xi_2 + \cdots + f_q(x)\xi_q$;
(c_2) constraint function $g_i(x, \xi) = g_{i0}(x) + g_{i1}(x)\xi_1 + g_{i2}(x)\xi_2 + \cdots + g_{iq}(x)\xi_q$,

where f_k and g_{ik} are nonnegative functions for all $1 \leq i \leq n$ and $0 \leq k \leq q$.

Theorem 5.8 *Assume that $\xi_1, \xi_2, \ldots, \xi_q$ are independent triangular fuzzy variables. If the entropy optimization model satisfies conditions $(c_1$–$c_2)$, then it has the following crisp equivalent,*

$$\begin{cases} \min & f\left(x, H[\xi]\right) - f_0(x) \\ \text{s.t.} & f\left(x, E[\xi]\right) \geq \alpha \\ & g_i\left(x, E[\xi]\right) \leq 0, \quad i = 1, 2, \ldots, n \end{cases} \tag{5.9}$$

where $H[\xi] = (H[\xi_1], H[\xi_2], \ldots, H[\xi_m])$ and $E[\xi] = (E[\xi_1], E[\xi_2], \ldots, E[\xi_m])$.

Proof Since fuzzy variables $\xi_1, \xi_2, \ldots, \xi_q$ are independent, for each feasible solution x, the objective value is a triangular fuzzy variable

$$f(x, \xi) = \left(f(x, a), f(x, b), f(x, c)\right),$$

which has the expected value

$$E\left[f(x, \xi)\right] = \left(f(x, a) + 2f(x, b) + f(x, c)\right)/4 = f\left(x, E[\xi]\right)$$

and has the entropy

$$H\left[f(x, \xi)\right] = \left(f(x, c) - f(x, a)\right)/2 = f\left(x, H[\xi]\right) - f_0(x).$$

Similarly, for each feasible solution x, the constraint function $g_i(x, \xi)$ is a triangular fuzzy variable with expected value

$$E\left[g_i(x, \xi)\right] = \left(g_i(x, a) + 2g_i(x, b) + g_i(x, c)\right)/4 = g_i\left(x, E[\xi]\right)$$

for each $i = 1, 2, \ldots, n$. The proof is complete. $\qquad\qquad\square$

Remark 5.3 Note that Theorem 5.8 also holds for equipossible fuzzy variables, trapezoidal fuzzy variables, and normal fuzzy variables.

In many cases, there are multiple conflicting objectives. If the decision-maker prefers to get a solution with the minimum uncertainty for all objectives, we have the following multi-objective entropy optimization model

$$\begin{cases} \min & \left[H[f_1(x, \xi)], H[f_2(x, \xi)], \ldots, H[f_p(x, \xi)]\right] \\ \text{s.t.} & E\left[f_j(x, \xi)\right] \geq \alpha_j, \quad j = 1, 2, \ldots, p \\ & E\left[g_i(x, \xi)\right] \leq 0, \quad i = 1, 2, \ldots, n. \end{cases}$$

5.4 Fuzzy Simulation

In order to solve the general entropy optimization models, this section introduces a fuzzy simulation technique to approximate the fuzzy entropy

$$U : x \rightarrow H\left[f(x, \xi)\right] \tag{5.10}$$

where f is a real function, and ξ is a fuzzy vector.

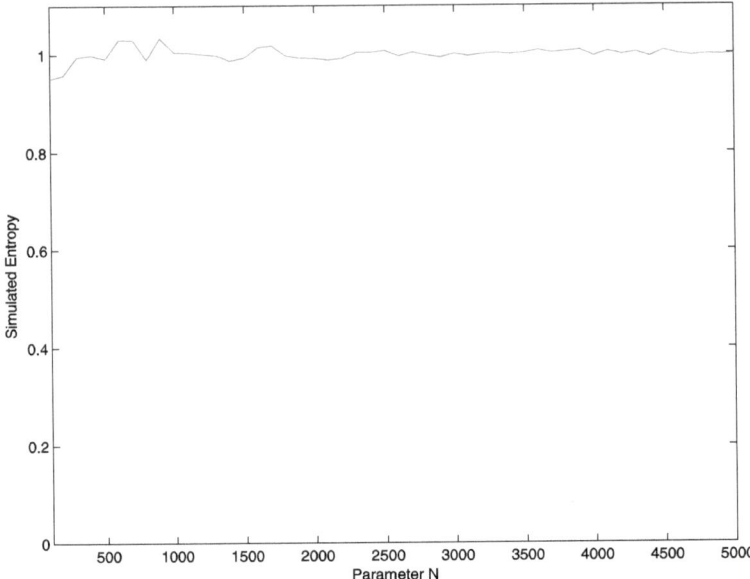

Fig. 5.2 Entropy simulation with variable parameter N

Suppose that $\boldsymbol{\xi}$ has a joint credibility function ν. Randomly generate vectors $\boldsymbol{y}_i$ with credibilities ν_i for all $i = 1, 2, \ldots, N$. According to Definition 5.2, the entropy can be simulated as the numerical integration of function $S(\nu(x))$. The procedure is listed as follows.

Algorithm 5.1 (Fuzzy simulation for entropy)

Step 1. Set $h = 0$ and $k = 0$.
Step 2. Randomly generate $\boldsymbol{y}_i$ with credibilities ν_i for all $i = 1, 2, \ldots, N$.
Step 3. Calculate the minimum and maximum values

$$a = \min\big\{ f(\boldsymbol{x}, \boldsymbol{y}_i) \mid 1 \le i \le N \big\}, \qquad b = \max\big\{ f(\boldsymbol{x}, \boldsymbol{y}_i) \mid 1 \le i \le N \big\}.$$

Step 4. Calculate $s_k = -\nu_k \ln \nu_k - (1 - \nu_k) \ln(1 - \nu_k)$.
Step 5. Set $h \to h + s_k$. If $k < N$, set $k = k + 1$ and go to step 4.
Step 6. Return $h(b - a)/N$ as the entropy.

Example 5.8 Taking $\xi = (1, 2, 3)$, we study the convergence of the entropy simulation algorithm by changing N from 100 to 5000 with step of 100. The results are illustrated by Fig. 5.2. It is shown that when $N \ge 2000$, the simulated results are stable and converge to the exact value $H[\xi] = 1$.

Example 5.9 Taking $N = 2000$, we perform Algorithm 5.1 on twenty fuzzy variables, including equipossible fuzzy variables, triangular fuzzy variables, trapezoidal

Table 5.1 Simulation results on entropy

Fuzzy variables	Simulated value	Exact value	Relative error
(0.0, 1.0)	0.6931	0.6931	0.0000
(−1.0, 2.0)	2.0794	2.0794	0.0000
(3.5, 7.8)	2.9803	2.9805	0.0001
(−50, −10)	27.7233	27.7259	0.0001
(−0.3, 1.8, 2.3)	1.3017	1.3000	0.0013
(1.5, 3.0, 4.1)	1.2833	1.3000	0.0128
(10, 15, 20)	4.9387	5.0000	0.0123
(25, 40, 50)	12.5234	12.5000	0.0019
(1.0, 2.0, 3.0, 4.0)	1.6843	1.6931	0.0052
(3.1, 4.2, 4.5, 6.0)	1.5002	1.5079	0.0051
(2.4, 3.6, 3.7, 5.6)	4.0464	4.0763	0.0073
(10, 25, 30, 45)	1.5867	1.6193	0.0201
N(1.5, 1.0)	2.5464	2.5651	0.0073
N(2.5, 1.0)	2.5688	2.5651	0.0014
N(3.7, 2.1)	5.4003	5.3867	0.0025
N(6.0, 1.3)	3.3310	3.3346	0.0011
E(1.3)	1.6476	1.6673	0.0118
E(1.0)	1.2907	1.2825	0.0063
E(3.5)	4.4391	4.4889	0.0111
E(7.6)	9.7563	9.7474	0.0009

fuzzy variables, exponential fuzzy variables, and normal fuzzy variables. We record the simulated results by Table 5.1, and make comparisons with the exact values. The last column records the relative error (3.25), which ranges from 0.00 % to 2.01 % and the average error is 0.54 %. These results imply that the simulation algorithm can obtain a very satisfactory approximation for entropy.

Example 5.10 In this example, we take $N = 2000$, and perform Algorithm 5.1 fifty times on the triangular fuzzy variable $\xi = (-0.3, 1.8, 2.3)$. The results are recorded by Table 5.2. Compared with the exact value $H[\xi] = 1.3$, the relative error ranges from 0.00 % to 0.50 %. This example tells us that the entropy simulation is a stochastic algorithm such that we may get different values if we perform the algorithm more than one times.

5.5 Applications

This section applies the entropy optimization model to study the fuzzy portfolio selection problem. See Examples 2.4 and 2.5.

Table 5.2 Entropy simulation on $\xi = (-0.3, 1.8, 2.3)$

	Simulated value	Error		Simulated value	Error
1	1.2962	0.0030	26	1.3002	0.0001
2	1.3016	0.0012	27	1.3006	0.0005
3	1.2951	0.0038	28	1.3009	0.0007
4	1.3013	0.0010	29	1.2994	0.0005
5	1.3047	0.0036	30	1.3004	0.0003
6	1.3036	0.0028	31	1.3026	0.0020
7	1.3014	0.0011	32	1.3029	0.0022
8	1.2989	0.0008	33	1.3049	0.0038
9	1.2938	0.0048	34	1.2946	0.0041
10	1.3033	0.0025	35	1.3017	0.0013
11	1.3043	0.0033	36	1.3016	0.0012
12	1.3015	0.0011	37	1.2990	0.0008
13	1.3025	0.0019	38	1.3010	0.0008
14	1.2992	0.0006	39	1.3002	0.0001
15	1.2984	0.0012	40	1.2974	0.0020
16	1.2944	0.0043	41	1.3043	0.0033
17	1.3004	0.0003	42	1.3016	0.0013
18	1.3013	0.0010	43	1.3026	0.0020
19	1.2993	0.0005	44	1.2992	0.0006
20	1.2995	0.0004	45	1.3007	0.0006
21	1.2936	0.0050	46	1.3023	0.0017
22	1.2991	0.0007	47	1.2987	0.0010
23	1.2994	0.0005	48	1.3012	0.0009
24	1.3015	0.0011	49	1.2963	0.0029
25	1.2955	0.0035	50	1.2961	0.0030

Example 5.11 Suppose that there are four stocks with independent normal fuzzy returns (see Table 3.3). If the investor would like to minimize the entropy with return level 1.4, then we get the following formulation

$$\begin{cases} \min & x_1 + 1.1x_2 + 1.3x_3 + 1.4x_4 \\ \text{s.t.} & 1.1x_1 + 1.3x_2 + 1.5x_3 + x_4 \geq 1.4 \\ & x_1 + x_2 + x_3 + x_4 = 1 \\ & x_1, x_2, x_3, x_4 \geq 0. \end{cases}$$

We use the Matlab function *Linprog* to solve the linear programming model. It is shown that the optimal portfolio is

$$x_1 = 0, \qquad x_2 = 0.5, \qquad x_3 = 0.5, \qquad x_4 = 0,$$

and the objective value is 1.2.

Example 5.12 In this example, the entropy optimization model is applied to the data shown in Table 3.4, which is composed of two triangular fuzzy variables and two normal fuzzy variables. With expected return level 1.4, if the decision-maker prefers a portfolio with a smaller entropy, we have the following entropy optimization model,

$$\begin{cases} \min & H[\xi_1 x_1 + \xi_2 x_2 + \xi_3 x_3 + \xi_4 x_4] \\ \text{s.t.} & E[\xi_1 x_1 + \xi_2 x_2 + \xi_3 x_3 + \xi_4 x_4] \geq 1.4 \\ & x_1 + x_2 + x_3 + x_4 = 1 \\ & x_1, x_2, x_3, x_4 \geq 0. \end{cases}$$

Take $N = 3000$, $G = 30$, $P_c = 0.4$, $P_m = 0.2$ and *pop-size* = 100. A run of the genetic algorithm shows that the optimal portfolio is

$$x_1 = 0.1271, \qquad x_2 = 0.1630, \qquad x_3 = 0.1042, \qquad x_4 = 0.6057,$$

and the entropy is 6.2707.

References

Bhandari D, Pal NR (1993) Some new information measures of fuzzy sets. Inf Sci 67:209–228

De Luca A, Termini S (1972) A definition of nonprobabilistic entropy in the setting of fuzzy sets theory. Inf Control 20:301–312

Kaufmann A (1975) Introduction to the theory of fuzzy subsets. Academic Press, New York

Kosko B (1986) Fuzzy entropy and conditioning. Inf Sci 40:165–174

Li X, Liu B (2007) Maximum entropy principle for fuzzy variables. Int J Uncertain Fuzziness Knowl-Based Syst 15(2):43–52

Li PK, Liu B (2008a) Entropy of credibility distributions for fuzzy variables. IEEE Trans Fuzzy Syst 16(1):123–129

Li X, Ralescu D, Tang T (2011) Fuzzy train energy consumption minimization model and algorithm. Iran J Fuzzy Syst 8(4):77–91

Liu XC (1992) Entropy, distance measure and similarity measure of fuzzy sets and their relations. Fuzzy Sets Syst 52:305–318

Pal NR, Bezdek JC (1994) Measuring fuzzy uncertainty. IEEE Trans Fuzzy Syst 2:107–118

Pal NR, Pal SK (1992) Higher order fuzzy entropy and hybrid entropy of a set. Inf Sci 61:211–231

Szmidt E, Kacprzyk J (2001) Entropy for intuitionistic fuzzy sets. Fuzzy Sets Syst 118:467–477

Yager RR (1979) On measures of fuzziness and negation, part I: membership in the unit interval. Int J Gen Syst 5:221–229

Chapter 6
Cross-Entropy Minimization Model

Cross-entropy is used to characterize the divergence between two fuzzy variables. In 1992, Kapur and Kesavan (1992) proposed a cross-entropy minimization model to minimize the divergence of the distribution on objective values from a priori distribution. From then on, many researchers accepted and investigated this new modeling method (Cherny and Maslov 2003; Fang et al. 1997; Qin et al. 2009; Rubinstein 2008; Simonelli 2005). Based on the concepts of fuzzy entropy, Bhandari and Pal (1993) defined the cross-entropy for fuzzy set by using membership function, and Li and Liu (2012) defined the cross-entropy for fuzzy variable by using credibility function. Furthermore, Qin et al. (2009) proposed a cross-entropy minimization model to study the fuzzy portfolio selection problems.

This chapter mainly includes the definition of cross-entropy, minimum cross-entropy principle, cross-entropy minimization model, fuzzy simulation and applications.

6.1 Cross-Entropy

This section defines a concept of fuzzy cross-entropy for quantifying the divergences of fuzzy variables from a priori one.

Definition 6.1 Let ξ and η be two discrete fuzzy variables taking values in $\{x_1, x_2, \ldots\}$. Then the fuzzy cross-entropy of ξ from η is defined as

$$D[\xi; \eta] = \sum_{i=1}^{\infty} T\left(\mathrm{Cr}\{\xi = x_i\}, \mathrm{Cr}\{\eta = x_i\}\right) \tag{6.1}$$

where $T : [0, 1] \times [0, 1] \to [0, \infty)$ is a binary function defined as

$$T(s, t) = s \ln(s/t) + (1 - s) \ln\big((1 - s)/(1 - t)\big)$$

X. Li, *Credibilistic Programming*, Uncertainty and Operations Research,
DOI 10.1007/978-3-642-36376-4_6, © Springer-Verlag Berlin Heidelberg 2013

Fig. 6.1 The shape of
function $T(s, t)$

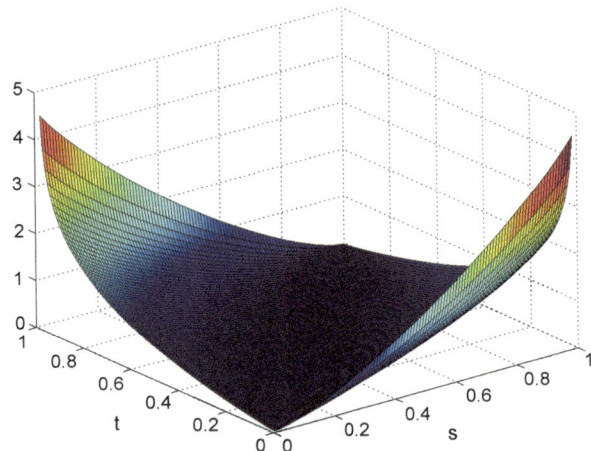

with boundary conditions

$$T(s,0) = \begin{cases} 0, & \text{if } s = 0 \\ +\infty, & \text{if } s > 0, \end{cases} \qquad T(s,1) = \begin{cases} 0, & \text{if } s = 1 \\ +\infty, & \text{if } s < 1. \end{cases}$$

If fuzzy variables ξ and η have credibility functions ν and μ, respectively, then
the cross-entropy of ξ from η is

$$D[\xi; \eta] = \sum_{i=1}^{\infty} T\big(\nu(x_i), \mu(x_i)\big). \tag{6.2}$$

It is easy to prove that function T has a gradient function

$$\nabla T(s,t) = \left(\ln\left(\frac{s}{t}\right) - \ln\left(\frac{1-s}{1-t}\right), \frac{1-s}{1-t} - \frac{s}{t} \right), \tag{6.3}$$

and has a Hessian matrix

$$H(s,t) = \begin{pmatrix} \dfrac{1}{s(1-s)} & -\dfrac{1}{t(1-t)} \\ -\dfrac{1}{t(1-t)} & \dfrac{s}{t^2} + \dfrac{1-s}{(1-t)^2} \end{pmatrix}. \tag{6.4}$$

Then the following properties about $T(s, t)$ can be easily proved (see Fig. 6.1):
(a) it is a strictly convex function with respect to (s, t) and attains its minimum
value zero when $s = t$; (b) for any $0 \leq s \leq 1$ and $0 \leq t \leq 1$, we have $T(s, t) = T(1-s, 1-t)$.

Remark 6.1 It is easy to prove that the cross-entropy is permutationally symmetric,
that is, the value does not change if the outcomes are labeled differently.

Remark 6.2 The concept of cross-entropy measures the divergence of a fuzzy variable from a priori one instead of the distance. Hence, the symmetry property is not necessary. That is, we may have $D[\xi; \eta] \neq D[\eta; \xi]$. For example, let ξ and η be two simple fuzzy variables with credibility functions

$$v(x) = \begin{cases} 0.3, & \text{if } x = x_1 \\ 0.4, & \text{if } x = x_2 \\ 0.6, & \text{if } x = x_3, \end{cases} \qquad \mu(x) = \begin{cases} 0.2, & \text{if } x = x_1 \\ 0.5, & \text{if } x = x_2 \\ 0.5, & \text{if } x = x_3, \end{cases}$$

respectively. First, it is easy to prove that

$$T(0.3, 0.2) = 0.0282, \qquad T(0.4, 0.5) = 0.0201, \qquad T(0.6, 0.5) = 0.0201.$$

Then according to Definition 6.1, we have the cross-entropy of ξ from η is $D[\xi; \eta] = 0.0684$. On the other hand, it follows from

$$T(0.2, 0.3) = 0.0257, \qquad T(0.5, 0.4) = 0.0204, \qquad T(0.5, 0.6) = 0.0204$$

that the cross-entropy of η from ξ is $D[\eta; \xi] = 0.0665$.

Definition 6.2 Let ξ and η be two continuous fuzzy variables. Then the cross-entropy of ξ from η is defined as

$$D[\xi; \eta] = \int_{-\infty}^{\infty} T(v(x), \mu(x)) \, dx$$

where v and μ are the credibility functions of ξ and η, respectively.

Example 6.1 Let $\xi = (a, b, c)$ be a triangular fuzzy variable with credibility function v, and let $\tau = (a, c)$ be an equipossible fuzzy variable. According to Definition 6.2, it is easy to prove that

$$D[\xi; \tau] = \int_a^c v(x) \ln(v(x)/0.5) + (1 - v(x)) \ln((1 - v(x))/0.5) \, dx$$

$$= \int_a^c v(x) \ln v(x) + (1 - v(x)) \ln(1 - v(x)) \, dx + \int_a^c \ln 2 \, dx$$

$$= (\ln 2 - 0.5)(c - a).$$

Now, we consider the cross-entropy of fuzzy variable τ from ξ. First, we have

$$D[\tau; \xi] = \int_a^c 0.5 \ln(0.5/v(x)) + 0.5 \ln(0.5/(1 - v(x))) \, dx$$

$$= \int_a^c -0.5 \ln v(x) - 0.5 \ln(1 - v(x)) \, dx - \int_a^c \ln 2 \, dx$$

$$= (a - c) \ln 2 - 0.5 \int_a^c \ln v(x) + \ln(1 - v(x)) \, dx.$$

Furthermore, it follows from the theorem of integration by parts that

$$\int_a^b \ln v(x) + \ln(1 - v(x)) \, dx$$

$$= \int_a^b \ln\big((x-a)/2(b-a)\big) + \ln\big(1 - (x-a)/2(b-a)\big) \, dx$$

$$= -2(b-a),$$

and

$$\int_b^c \ln v(x) + \ln(1 - v(x)) \, dx$$

$$= \int_b^c \ln\big((c-x)/2(c-b)\big) + \ln\big(1 - (c-x)/2(c-b)\big) \, dx$$

$$= -2(c-b),$$

which implies that the cross-entropy of τ from ξ is

$$D[\tau; \xi] = (1 - \ln 2)(c - a).$$

Example 6.2 Let $\xi = (a, b, c, d)$ be a trapezoidal fuzzy variable with credibility function v, and let $\tau = (a, c)$ be an equipossible fuzzy variable. It is easy to prove that

$$\int_a^b T\big(v(x), 0.5\big) \, dx = (\ln 2 - 0.5)(b - a),$$

$$\int_b^c T\big(v(x), 0.5\big) \, dx = \int_b^c 0.5 \ln(0.5/0.5) + 0.5 \ln(0.5/0.5) \, dx = 0,$$

$$\int_c^d T\big(v(x), 0.5\big) \, dx = (\ln 2 - 0.5)(d - c).$$

According to Definition 6.2, the cross-entropy of ξ from τ is

$$D[\eta; \tau] = (\ln 2 - 0.5)(b - a) + (\ln 2 - 0.5)(d - c)$$

$$= (\ln 2 - 0.5)(d - c + b - a).$$

Now, we consider the cross-entropy of fuzzy variable τ from ξ. First, we have

$$D[\tau; \xi] = \int_a^d 0.5 \ln\big(0.5/v(x)\big) + 0.5 \ln\big(0.5/(1 - v(x))\big) \, dx$$

$$= \int_a^d -0.5 \ln v(x) - 0.5 \ln\big(1 - v(x)\big) \, dx - \int_a^d \ln 2 \, dx$$

$$= (a - d) \ln 2 - 0.5 \int_a^d \ln v(x) + \ln\big(1 - v(x)\big) \, dx.$$

Furthermore, it follows from the theorem of integration by parts that

$$\int_a^b \ln \nu(x) + \ln\big(1 - \nu(x)\big)\,dx = -2(b - a),$$

$$\int_b^c \ln \nu(x) + \ln\big(1 - \nu(x)\big)\,dx = \int_b^c \ln 0.5 + \ln 0.5\,dx = 2(c - b)\ln 0.5,$$

$$\int_c^d \ln \nu(x) + \ln\big(1 - \nu(x)\big)\,dx = -2(d - c),$$

which implies that

$$D[\tau; \xi] = (d - c + b - a)(1 - \ln 2).$$

Theorem 6.1 *For any fuzzy variables ξ and η, we have*

$$D[\xi; \eta] \geq 0$$

and the equality holds if and only if ξ and η are identically distributed.

Proof Let ν and μ be the credibility functions of fuzzy variables ξ and η, respectively. It follows from the nonnegativity of function $T(s, t)$ that

$$D[\xi; \eta] = \int_{-\infty}^{\infty} T\big(\nu(x), \mu(x)\big)\,dx \geq 0.$$

Furthermore, the equality holds if and only if $T(\nu(x), \mu(x)) = 0$, i.e., $\nu(x) = \mu(x)$ for all $x \in \Re$, which implies that ξ and η are identically distributed. If ξ and η are discrete fuzzy variables, the theorem can be proved in a similar way. The proof is complete. □

Theorem 6.2 *Let ξ and η be two continuous fuzzy variables. For any real numbers a and b, we have*

$$D[a\xi + b; a\eta + b] = |a|D[\xi; \eta].$$

Proof The conclusion is trivial when $a = 0$. In what follows, we assume that $a \neq 0$. Let ν and μ be the credibility functions of fuzzy variables ξ and η, respectively. It follows from Definition 6.2 that

$$D[a\xi + b; a\eta + b] = \int_{-\infty}^{\infty} T\big(\nu\big((x - b)/a\big), \mu\big((x - b)/a\big)\big)\,dx$$

$$= \int_{-\infty}^{\infty} |a|T\big(\nu(t), \mu(t)\big)\,dt$$

$$= |a|D[\xi; \eta].$$

The proof is complete. □

Theorem 6.3 *Let* $\tau = (a, b)$ *be an equipossible fuzzy variable. Then for any continuous fuzzy variable* ξ *taking values in* $[a, b]$, *we have*

$$D[\xi, \tau] = H[\tau] - H[\xi].$$

Proof Let ν be the credibility function of ξ. It follows from the definition of cross-entropy that

$$D[\xi, \tau] = \int_a^b \nu(x) \ln\big(2\nu(x)\big) + \big(1 - \nu(x)\big) \ln\big(2 - 2\nu(x)\big) dx$$

$$= \int_a^b \ln 2 + \nu(x) \ln \nu(x) + \big(1 - \nu(x)\big) \ln\big(1 - \nu(x)\big) dx$$

$$= (b - a) \ln 2 - H[\xi]$$

$$= H[\tau] - H[\xi].$$

The proof is complete.   □

Theorem 6.4 *Let* τ *be an equipossible fuzzy variable taking values in* $\{x_1, x_2, \ldots, x_n\}$. *For any fuzzy variable* ξ *taking values in* $\{x_1, x_2, \ldots, x_n\}$, *we have*

$$D[\xi, \tau] = H[\tau] - H[\xi].$$

Proof It may be proved similarly with Theorem 6.3.   □

6.2 Minimum Cross-Entropy Principle

In many real problems, the credibility function of a fuzzy variable is unavailable except some partial information, for example, moment constraints, which may be based on observations. In this case, the maximum entropy principle tells us that out of all the credibility functions satisfying given constraints, choose the one that has maximum entropy. However, there may be another type of information, for example, a prior credibility function, which may be based on intuition or experience with the problem. If both the a prior credibility function and the moment constraints are given, which credibility function should we choose? The following *minimum cross-entropy principle* tells us that: *out of all credibility functions satisfying given moment constraints, choose the one that is closest to the given a priori credibility function.*

There is nothing mysterious about this principle. It is just based on common sense. Our credibility function must be consistent with observations, and if there are many credibility functions consistent with these observations, we choose the one that is nearest to the intuition and experience.

Remark 6.3 If there is no a priori experience or intuition to guide us, we choose the credibility function that is nearest to the equipossible one τ. In this sense, the maximum entropy principle and the minimum cross-entropy principle are consistent because

$$H[\xi] = H[\tau] - D[\xi; \tau].$$

6.3 Cross-Entropy Minimization Model

For the fuzzy programming problems, it is possible that some decision-makers could provide a priori credibility function for the fuzzy objective. In this case, Qin et al. (2009) proposed a cross-entropy minimization model, which minimizes the divergence from the prior fuzzy variable η under certain expected constraints

$$\begin{cases} \min & D\big[f(x, \xi); \eta\big] \\ \text{s.t.} & E\big[g_i(x, \xi)\big] \leq 0, \quad i = 1, 2, \ldots, n. \end{cases} \tag{6.5}$$

Remark 6.4 The concepts of feasible solution, local optimal solution, and global optimal solution are given by Definitions 2.6, 2.7, and 2.8.

In many cases, there are multiple conflicting objectives. If the decision-maker prefers to get a solution with the minimum cross-entropy for all objectives, we have the following multi-objective cross-entropy minimization model

$$\begin{cases} \min & \big[D\big[f_1(x, \xi); \eta_1\big], D\big[f_2(x, \xi); \eta_2\big], \ldots, D\big[f_p(x, \xi); \eta_p\big]\big] \\ \text{s.t.} & E\big[g_i(x, \xi)\big] \leq 0, \quad i = 1, 2, \ldots, n. \end{cases}$$

Remark 6.5 The expected value constraints can be changed to chance constraints if the decision-maker prefers to ensure the feasibility of solution with certain credibilities.

6.4 Fuzzy Simulation

Generally speaking, it is difficult to calculate the cross-entropy analytically. This section introduces a simulation method to approximate the cross-entropy

$$U : x \rightarrow D\big[f(x, \xi); \eta\big]. \tag{6.6}$$

For each feasible solution x, suppose that $f(x, \xi)$ has a credibility function v, and fuzzy variable η has a credibility function μ. According to Definition 6.2, the cross-entropy can be simulated as the numerical integration of function

$$T(v, \mu) = v \ln(v/\mu) + (1 - v) \ln\big((1 - v)/(1 - \mu)\big).$$

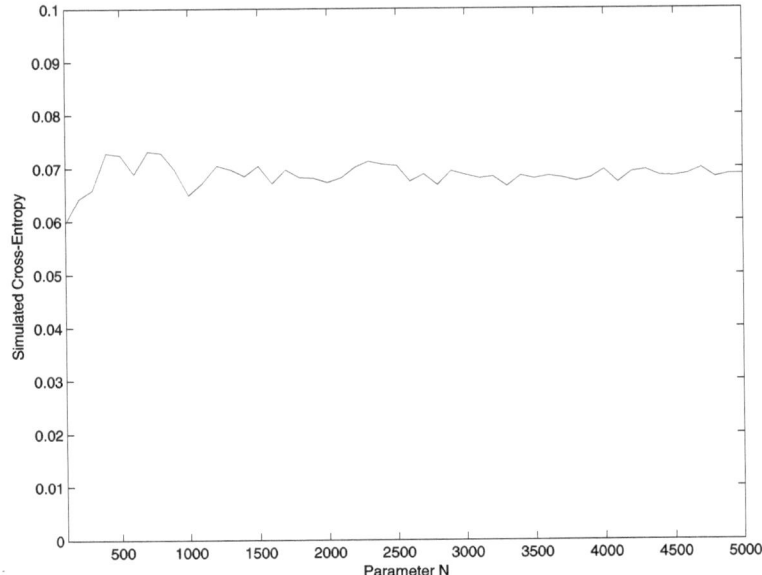

Fig. 6.2 The cross-entropy simulation with variable parameter N

Randomly generate vectors y_i and calculate the objective values $f_i = f(x, y_i)$ for all $i = 1, 2, \ldots, N$. Furthermore, calculate the credibilities v_i and μ_i for all $i = 1, 2, \ldots, N$. Then the cross-entropy can be simulated as

$$D\big[f(x, \xi); \eta\big] = \sum_{i=1}^{N} T(v_i, \mu_i)(b - a)/N$$

provided that N is sufficient large, where

$$a = \min\big\{f(x, y_i) \mid 1 \le i \le N\big\}, \qquad b = \max\big\{f(x, y_i) \mid 1 \le i \le N\big\}.$$

The procedure is listed as follows.

Algorithm 6.1 (Fuzzy simulation for cross-entropy)

Step 1. Set $c = 0$ and $k = 0$.
Step 2. Randomly generate y_i for all $i = 1, 2, \ldots, N$.
Step 3. Calculate the objective values $f_i = f(x, y_i)$ and the credibilities

$$\mu_i = \mathrm{Cr}\{\eta = f_i\}, \quad i = 1, 2, \ldots, N,$$
$$v_i = \mathrm{Cr}\big\{f(x, \xi) = f_i\big\}, \quad i = 1, 2, \ldots, N.$$

Step 4. Calculate the minimum and maximum objective values a and b.
Step 5. Calculate $t_k = v_k \ln(v_k/\mu_k) + (1 - v_k) \ln((1 - v_k)/(1 - \mu_k))$.

Table 6.1 Simulation results on fuzzy cross-entropy

Fuzzy variable ξ	Fuzzy variable η	$D[\xi; \eta]$	$D[\eta; \xi]$
(0.0, 1.0)	(0.0, 0.5, 1.0)	0.3112	0.1925
(−1.0, 2.0)	(−1.0, 0.0, 1.0, 2.0)	0.6204	0.3931
(3.5, 7.8)	(3.5, 4.0, 7.8)	1.3214	0.8387
(−50.0, −10.0)	(−50.0, −30.0, −10.0)	12.1506	7.5892
(−0.3, 1.8, 2.3)	(−0.3, 2.3)	0.5068	0.7956
(1.5, 3.0, 4.1)	(1.5, 3.7, 4.1)	0.1092	0.1215
(10.0, 15.0, 20.0)	(10.0, 18.0, 20.0)	0.4653	0.5006
(25.0, 40.0, 50.0)	(25.0, 50.0)	4.8068	7.5264
(1.0, 2.0, 3.0, 4.0)	(1.0, 1.3, 3.8, 4.0)	0.2156	0.2749
(3.1, 4.2, 4.5, 6.0)	(3.1, 6.0)	0.4986	0.8060
(2.4, 3.6, 3.7, 5.6)	(2.4, 4.6, 4.7, 5.6)	0.1661	0.1718
(10.0, 25.0, 30.0, 45.0)	(10.0, 15.0, 45.0)	2.5277	2.5976
N(1.5, 1.0)	N(2.0, 1.7)	0.2007	0.2624
N(2.5, 1.0)	N(4.0, 1.0)	1.0895	1.0151
N(3.7, 2.1)	N(0.0, 1.5)	4.0891	2.9882
N(6.0, 1.3)	N(5.0, 3.0)	0.5707	0.8555
E(1.3)	E(3.1)	0.1416	0.2188
E(1.0)	E(3.2)	0.2336	0.3873
E(3.5)	E(2.0)	0.0948	0.0703
E(7.6)	E(10.0)	0.0333	0.0388

Step 6. Set $c \to c + t_k$. If $k < N$, set $k = k + 1$ and go to step 5.
Step 7. Return $c(b - a)/N$ as the value of cross-entropy.

Example 6.3 Taking triangular fuzzy variables $\xi = (1, 2, 3)$ and $\eta = (1, 1.5, 3)$ for example, we study the convergence of the cross-entropy simulation algorithm. We change N from 100 to 5000 with step of 100, and illustrate the simulated results by Fig. 6.2. It is shown that the algorithm converges when N is larger than 3000.

Example 6.4 This example applies the simulation algorithm with $N = 3000$ to approximate the cross-entropy for twenty pairs of fuzzy variables. In Table 6.1, each line records one pair of simulated results, where the third column is the cross-entropy of ξ from η, and the last column is the cross-entropy of η from ξ. Taking the first line for example, there are an equipossible fuzzy variable $\xi = (0, 1)$ and a triangular fuzzy variable $\eta = (0, 0.5, 1)$. The cross-entropy of ξ from η is 0.3112, and the cross-entropy of η from ξ is 0.1925.

Example 6.5 Since the cross-entropy simulation is a stochastic algorithm, the simulated results obtained from different performances are generally different. In this example, we take $N = 3000$, and perform Algorithm 6.1 fifty times on the triangular

Table 6.2 Cross-entropy simulation of $(-0.3, 1.8, 2.3)$ from $(-0.3, 2.3)$

No.	Simulated value	Error	No.	Simulated value	Error
1	0.5030	0.0016	26	0.4997	0.0050
2	0.5025	0.0005	27	0.5015	0.0014
3	0.5035	0.0025	28	0.5006	0.0031
4	0.5021	0.0003	29	0.5037	0.0029
5	0.5003	0.0037	30	0.5025	0.0007
6	0.5008	0.0028	31	0.5026	0.0008
7	0.5006	0.0031	32	0.5003	0.0037
8	0.5027	0.0011	33	0.5008	0.0028
9	0.5023	0.0002	34	0.5022	0.0000
10	0.5033	0.0022	35	0.4998	0.0048
11	0.5019	0.0005	36	0.5033	0.0021
12	0.5023	0.0002	37	0.5025	0.0006
13	0.5006	0.0031	38	0.4997	0.0048
14	0.5041	0.0038	39	0.5025	0.0006
15	0.5040	0.0037	40	0.5023	0.0002
16	0.5032	0.0021	41	0.5012	0.0019
17	0.5004	0.0035	42	0.5021	0.0002
18	0.5024	0.0005	43	0.5006	0.0032
19	0.5005	0.0033	44	0.5036	0.0027
20	0.5025	0.0007	45	0.5001	0.0041
21	0.5016	0.0011	46	0.5011	0.0021
22	0.5034	0.0025	47	0.5031	0.0019
23	0.5038	0.0032	48	0.5045	0.0046
24	0.5035	0.0027	49	0.5020	0.0004
25	0.5023	0.0002	50	0.5011	0.0021

fuzzy variable $\xi = (-0.3, 1.8, 2.3)$ and equipossible fuzzy variable $\eta = (-0.3, 2.3)$. The results are recorded by Table 6.2. We also show the relative error between the simulated value and the exact value 0.5022. It is shown that the relative error ranges from 0.00 % to 0.50 %, and the average value is 0.21 %.

6.5 Applications

This section applies the cross-entropy minimization model to study the fuzzy portfolio selection problem. See Examples 2.4 and 2.5. Suppose that there are four stocks with fuzzy returns. See Table 3.4. With the expected return level 1.4, if the decision-maker prefers a portfolio with a smaller divergence from a prior normal fuzzy return $\eta = N(1.5, 0.8)$, we have the following cross-entropy minimization model,

$$\begin{cases} \min & D[\xi_1 x_1 + \xi_2 x_2 + \xi_3 x_3 + \xi_4 x_4; \eta] \\ \text{s.t.} & E[\xi_1 x_1 + \xi_2 x_2 + \xi_3 x_3 + \xi_4 x_4] \geq 1.4 \\ & x_1 + x_2 + x_3 + x_4 = 1 \\ & x_1, x_2, x_3, x_4 \geq 0. \end{cases}$$

Take $N = 3000$, $G = 30$, $P_c = 0.4$, $P_m = 0.2$ and *pop-size* $= 100$. A run of the genetic algorithm shows that the optimal portfolio is

$$x_1 = 0.1609, \qquad x_2 = 0.0864, \qquad x_3 = 0.3635, \qquad x_4 = 0.3892,$$

and the minimum cross-entropy is 1.9066.

References

Bhandari D, Pal NR (1993) Some new information measures of fuzzy sets. Inf Sci 67:209–228

Cherny AS, Maslov VP (2003) On minimization and maximization of entropy in various disciplines. Theory Probab Appl 48:447–464

Fang S, Rajasekera J, Tsao H (1997) Entropy optimization and mathematical programming. Kluwer Academic, Boston

Kapur J, Kesavan H (1992) Entropy optimization principles with applications. Academic Press, New York

Li X, Liu B (2012) Fuzzy cross-entropy and its applications. Technical report

Qin ZF, Li X, Ji XY (2009) Portfolio selection based on fuzzy cross-entropy. J Comput Appl Math 228:139–149

Rubinstein R (2008) Semi-iterative minimum cross-entropy algorithms for rare-vents, counting, combinatorial and integer programming. Methodol Comput Appl Probab 10:121–178

Simonelli MR (2005) Indeterminacy in portfolio selection. Eur J Oper Res 163:170–176

Chapter 7
Regret Minimization Model

Distance between fuzzy quantities, used to represent the degree of difference, is a powerful concept in many disciplines of science and engineering. In 1983, Puri and Ralescu (1983) proposed the first Hausdorff-like distance, which takes the supremum of the Hausdorff distances between corresponding level sets. From then on, Hausdorff-like distance was studied and developed by many researchers such as Boxer (1997), Chaudhuri and Rosenfeld (1996, 1999), Diamond and Kloeden (1990), Fan (1998), Klement et al. (1986), and Rosenfeld (1985). Within the framework of credibility theory, Liu (2004) gave an Euclidean distance based on the concept of expected value. Furthermore, Li and Liu (2008b) proved the triangle inequality and the completeness of the fuzzy metric space. Based on the worst regret criterion (Inuiguchi and Ramík 2000; Inuiguchi and Tanino 2000), Li et al. (2012) proposed a fuzzy regret minimization model to minimize the distance between the fuzzy objective values and the best values.

This chapter mainly introduces the concept of distance, regret minimization model, and applications in the portfolio selection problem.

7.1 Distance

Fuzzy distance is used to measure the dissimilarity or difference between two fuzzy variables. Generally speaking, a distance function should satisfy the nonnegativity, identification, symmetry and triangle inequality.

Definition 7.1 (Liu 2004) The distance between fuzzy variables ξ and η is defined as

$$d[\xi, \eta] = E\big[|\xi - \eta|\big]. \tag{7.1}$$

X. Li, *Credibilistic Programming*, Uncertainty and Operations Research,
DOI 10.1007/978-3-642-36376-4_7, © Springer-Verlag Berlin Heidelberg 2013

Example 7.1 Take $(\Theta, \mathcal{A}, \mathrm{Cr})$ to be a credibility space $\{\theta_1, \theta_2\}$ with $\mathrm{Cr}\{\theta_1\} = 0.4$ and $\mathrm{Cr}\{\theta_2\} = 0.6$. Define fuzzy variables

$$\xi(\theta) = \begin{cases} 1, & \text{if } \theta = \theta_1 \\ 0, & \text{if } \theta = \theta_2, \end{cases} \qquad \eta(\theta) = \begin{cases} 2, & \text{if } \theta = \theta_1 \\ -2, & \text{if } \theta = \theta_2. \end{cases} \tag{7.2}$$

Since the absolute difference between ξ and η is

$$|\xi - \eta|(\theta) = \begin{cases} 1, & \text{if } \theta = \theta_1 \\ 2, & \text{if } \theta = \theta_2, \end{cases} \tag{7.3}$$

it follows from Example 3.2 that the distance is $d(\xi, \eta) = 0.4 \times 1 + 0.6 \times 2 = 1.6$.

Example 7.2 Suppose that $\xi = (a_1, b_1)$ and $\eta = (a_2, b_2)$ are two independent equipossible fuzzy variables. Then it has been proved that both $\xi - \eta$ and $\eta - \xi$ are equipossible fuzzy variables with

$$\xi - \eta = (a_1 - b_2, b_1 - a_2), \qquad \eta - \xi = (a_2 - b_1, b_2 - a_1).$$

If $a_1 > b_2$, then $\xi - \eta$ is a positive fuzzy variable such that

$$d(\xi, \eta) = E[\xi - \eta] = (a_1 - b_2 + b_1 - a_2)/2.$$

If $b_1 < a_2$, then $\xi - \eta$ is a negative fuzzy variable such that

$$d(\xi, \eta) = E[\eta - \xi] = (b_2 - a_1 + a_2 - b_1)/2.$$

Otherwise, we have $a_1 - b_2 \le 0 \le b_1 - a_2$. In this case, $|\xi - \eta|$ is an equipossible fuzzy variable taking values in $[0, (b_2 - a_1) \vee (b_1 - a_2)]$, and the distance is

$$d(\xi, \eta) = \big((b_2 - a_1) \vee (b_1 - a_2)\big)/2.$$

In general, the distance between ξ and η is

$$d(\xi, \eta) = \begin{cases} (a_1 - b_2 + b_1 - a_2)/2, & \text{if } a_1 > b_2 \\ (b_2 - a_1 + a_2 - b_1)/2, & \text{if } b_1 < a_2 \\ ((b_2 - a_1) \vee (b_1 - a_2))/2, & \text{otherwise.} \end{cases}$$

Example 7.3 Let $\xi = (a_1, b_1, c_1)$ and $\eta = (a_2, b_2, c_2)$ be two independent triangular fuzzy variables such that $(a_1, c_1) \cap (a_2, c_2) = \emptyset$ (see Fig. 7.1). First, it follows from the independence that $\xi - \eta$ is also a triangular fuzzy variable $(a_1 - c_2, b_1 - b_2, c_1 - a_2)$. If $c_1 \le a_2$, then $\xi - \eta$ is a nonpositive variable, and it follows from the Zadeh extension theorem that

$$|\xi - \eta| = (a_2 - c_1, b_2 - b_1, c_2 - a_1).$$

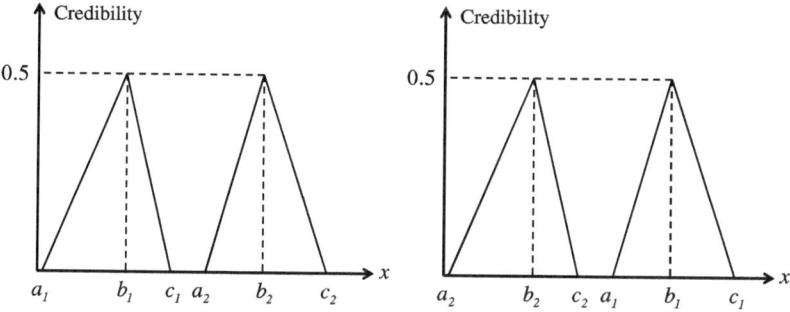

Fig. 7.1 Triangular fuzzy variables with $(a_1, c_1) \cap (a_2, c_2) = \emptyset$

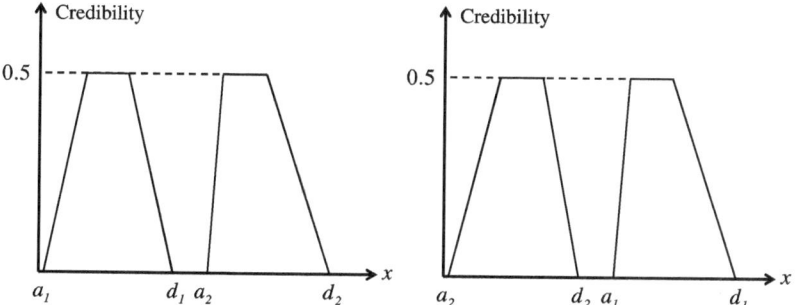

Fig. 7.2 Two trapezoidal fuzzy variables with $(a_1, d_1) \cap (a_2, d_2) = \emptyset$

According to Definition 7.1, we have

$$d(\xi, \eta) = \big((a_2 - c_1) + 2(b_2 - b_1) + (c_2 - a_1)\big)/4.$$

Similarly, if $c_2 \leq a_1$, then $\xi - \eta$ is a nonnegative variable such that

$$|\xi - \eta| = (a_1 - c_2, b_1 - b_2, c_1 - a_2).$$

According to Definition 7.1, the distance between ξ and η is

$$d(\xi, \eta) = \big((a_1 - c_2) + 2(b_1 - b_2) + (c_1 - a_2)\big)/4.$$

In general, the distance between triangular fuzzy variables ξ and η is

$$d(\xi, \eta) = \big(|a_1 - c_2| + 2|b_1 - b_2| + |c_1 - a_2|\big)/4.$$

Example 7.4 Suppose that $\xi = (a_1, b_1, c_1, d_1)$ and $\eta = (a_2, b_2, c_2, d_2)$ are two independent trapezoidal fuzzy variables such that $(a_1, d_1) \cap (a_2, d_2) = \emptyset$ (see Fig. 7.2). It has been proved that $\xi - \eta$ is also a trapezoidal fuzzy variable, denoted by $(a_1 - d_2, b_1 - c_2, c_1 - b_2, d_1 - a_2)$. The argument breaks down into two cases.

If $d_1 \le a_2$, then $\xi - \eta$ is a nonpositive variable, and it follows from the Zadeh extension theorem that

$$|\xi - \eta| = (a_2 - d_1, b_2 - c_1, c_2 - b_1, d_2 - a_1).$$

According to Definition 7.1, we have

$$d(\xi, \eta) = \big((a_2 - d_1) + (b_2 - c_1) + (c_2 - b_1) + (d_2 - a_1)\big)/4.$$

Similarly, if $d_2 \le a_1$, then $\xi - \eta$ is a nonnegative variable such that

$$|\xi - \eta| = (a_1 - d_2, b_1 - c_2, c_1 - b_2, d_1 - a_2).$$

According to Definition 7.1, the distance between ξ and η is

$$d(\xi, \eta) = \big((a_1 - d_2) + (b_1 - c_2) + (c_1 - b_2) + (d_1 - a_2)\big)/4.$$

In general, the distance between trapezoidal fuzzy variables ξ and η is

$$d(\xi, \eta) = \big(|a_1 - d_2| + |b_1 - c_2| + |c_1 - b_2| + |d_1 - a_2|\big)/4.$$

Example 7.5 Suppose that normal fuzzy variables $\xi = N(e_1, \sigma_1)$ and $\eta = N(e_2, \sigma_2)$ are mutually independent. Then $\xi - \eta$ is also a normal fuzzy variable with expected value $e = e_1 - e_2$ and standard variance $\sigma = \sigma_1 + \sigma_2$. First, we assume $e_1 \ge e_2$. It follows from the Zadeh extension theorem that fuzzy variable $|\xi - \eta|$ has the credibility function

$$v(x) = 1/\big(1 + \exp(\pi |x - e|/\sqrt{6}\sigma)\big), \quad x \ge 0.$$

Then it follows from the equations

$$\int_e^\infty v(x)\,dx = \frac{\sqrt{6}\sigma}{\pi}\ln 2,$$

$$\int_0^e v(x)\,dx = e + \frac{\sqrt{6}\sigma}{\pi}\big(\ln 2 - \ln(1 + \exp(\pi e/\sqrt{6}\sigma))\big)$$

that the distance between fuzzy variables ξ and η is

$$d(\xi, \eta) = \int_0^\infty \mathrm{Cr}\{|\xi - \eta| \ge r\}\,dr$$

$$= \int_0^e (1 - v(x))\,dx + \int_e^\infty v(x)\,dx$$

$$= \frac{\sqrt{6}\sigma}{\pi}\ln(1 + \exp(\pi e/\sqrt{6}\sigma)).$$

Similarly, if $e_1 < e_2$, we have

$$d(\xi, \eta) = \frac{\sqrt{6}\sigma}{\pi} \ln\bigl(1 + \exp(-\pi e/\sqrt{6}\sigma)\bigr).$$

In general, the distance between two normal fuzzy variables is

$$d(\xi, \eta) = \frac{\sqrt{6}(\sigma_1 + \sigma_2)}{\pi} \ln\bigl(1 + \exp(\pi|e_1 - e_2|/\sqrt{6}(\sigma_1 + \sigma_2))\bigr).$$

Example 7.6 Suppose that $\xi = EXP(m_1)$ and $\eta = EXP(m_2)$ are two exponential fuzzy variables with credibility functions v_1 and v_2, respectively. Denote μ as the credibility function of $|\xi - \eta|$. The argument breaks down into two cases. If $m_1 \geq m_2$, we have $v_1(x) \geq v_2(x)$ for all $x \geq 0$. It follows from the Zadeh extension theorem that

$$\begin{aligned}
\mu(x) &= \sup_{|y_1 - y_2| = x} \bigl(v_1(y_1) \wedge v_2(y_2)\bigr) \\[4pt]
&= \sup_{y_1 \geq 0, y_2 \geq 0, y_1 - y_2 = x} \bigl(v_1(y_1) \wedge v_2(y_2)\bigr) \\[4pt]
&= v_1(x) \wedge v_2(0) \\[4pt]
&= v_1(x)
\end{aligned}$$

for any $x \geq 0$. According to Definition 7.1, we have

$$d(\xi, \eta) = \int_0^\infty v_1(x)\, dx = \frac{\sqrt{6}\ln 2}{\pi} m_1.$$

Similarly, if $m_1 < m_2$, we can prove that

$$d(\xi, \eta) = \frac{\sqrt{6}\ln 2}{\pi} m_2.$$

In general, the distance between two exponential fuzzy variables is

$$d(\xi, \eta) = \frac{\sqrt{6}\ln 2}{\pi} (m_1 \vee m_2).$$

Theorem 7.1 (Li and Liu 2008b) *For any fuzzy variables ξ, η and τ, we have*

(a) *(Nonnegativity)* $d(\xi, \eta) \geq 0$;
(b) *(Identification)* $d(\xi, \eta) = 0$ *if and only if* $\xi = \eta$;
(c) *(Symmetry)* $d(\xi, \eta) = d(\eta, \xi)$;
(d) *(Triangle Inequality)* $d(\xi, \eta) \leq 2d(\xi, \tau) + 2d(\tau, \eta)$.

Proof The nonnegativity and symmetry follow immediately from the definition.
(b) If $\xi = \eta$, it is obvious that $d(\xi, \eta) = 0$. Conversely, suppose that $d(\xi, \eta) = 0$.
If $\xi \neq \eta$, then there is a point θ with $\text{Cr}\{\theta\} > 0$ such that $\xi(\theta) \neq \eta(\theta)$. Thus we have

$$d(\xi, \eta) = E\big[|\xi - \eta|\big] \geq \text{Cr}\{\theta\}\big|\xi(\theta) - \eta(\theta)\big| > 0.$$

The contradiction proves that $\xi = \eta$. Now we prove the part (d). In fact, for any
positive numbers a and b with $a + b = 1$, it follows from the credibility subadditivity
theorem that

$$d(\xi, \eta) = \int_0^\infty \text{Cr}\{|\xi - \eta| \geq r\}\mathrm{d}r \leq \int_0^\infty \text{Cr}\{|\xi - \tau| + |\tau - \eta| \geq r\}\mathrm{d}r$$

$$\leq \int_0^\infty \text{Cr}\big\{\{|\xi - \tau| \geq ar\} \cup \{|\tau - \eta| \geq br\}\big\}\mathrm{d}r$$

$$\leq \int_0^\infty \big(\text{Cr}\{|\xi - \tau| \geq ar\} + \text{Cr}\{|\tau - \eta| \geq br\}\big)\mathrm{d}r$$

$$= \int_0^\infty \text{Cr}\{|\xi - \tau| \geq ar\}\mathrm{d}r + \int_0^\infty \text{Cr}\{|\tau - \eta| \geq br\}\mathrm{d}r$$

$$= \frac{E[|\xi - \tau|]}{a} + \frac{E[|\tau - \eta|]}{b}$$

$$= \frac{d(\xi, \tau)}{a} + \frac{d(\tau, \eta)}{b}.$$

Especially, if we set $a = b = 0.5$, then we obtain $d(\xi, \eta) \leq 2d(\xi, \tau) + 2d(\tau, \eta)$. The
proof is complete. $\square$

Definition 7.2 (Li and Liu 2008b) Let $\mathbb{F}$ be the set of fuzzy variables. The set $\mathbb{F}$
with distance d is called a metric space of fuzzy variables, and is denoted by $(\mathbb{F}, d)$.

Theorem 7.2 (Li and Liu 2008b) *Metric space* $(\mathbb{F}, d)$ *is complete. That is, if*
$\xi_1, \xi_2, \ldots$ *are fuzzy variables on credibility space* $(\Theta, \mathcal{A}, \text{Cr})$ *and*

$$\lim_{i,j \to \infty} d(\xi_i, \xi_j) = 0, \tag{7.4}$$

then there is a fuzzy variable ξ *on* $(\Theta, \mathcal{A}, \text{Cr})$ *such that*

$$\lim_{i \to \infty} d(\xi_i, \xi) = 0. \tag{7.5}$$

Proof For any $\theta \in \Theta$ with nonzero credibility value, since

$$\lim_{i,j \to \infty} \big|\xi_i(\theta) - \xi_j(\theta)\big| \leq \lim_{i,j \to \infty} \frac{E[|\xi_i - \xi_j|]}{\text{Cr}\{\theta\}} = 0,$$

we know that $\{\xi_i(\theta)\}$ is a Cauchy sequence. Let $\xi(\theta)$ be the limit such that

$$\lim_{i \to \infty} \xi_i(\theta) = \xi(\theta).$$

It is clear that ξ is a fuzzy variable. For any $k \geq 1$, let i_k be the integer such that $E[|\xi_i - \xi_j|] \leq 1/4^k$ for any $i, j \geq i_k$. Without loss of generality, we assume that $i_1 < i_2 < \cdots < i_k < \cdots$. For any $k \geq 1$, it follows from the credibility subadditivity theorem that

$$E\left[|\xi_{i_k} - \xi|\right] = E\left[\left|\sum_{l \geq k}(\xi_{i_l} - \xi_{i_{l+1}})\right|\right] \leq E\left[\sum_{l \geq k}|\xi_{i_l} - \xi_{i_{l+1}}|\right]$$

$$\leq \int_0^{+\infty} \mathrm{Cr}\left\{\bigcup_{l \geq k}|\xi_{i_l} - \xi_{i_{l+1}}| \geq \frac{r}{2^l}\right\}\mathrm{d}r$$

$$\leq \int_0^{+\infty} \sum_{l \geq k}\mathrm{Cr}\left\{|\xi_{i_l} - \xi_{i_{l+1}}| \geq \frac{r}{2^l}\right\}\mathrm{d}r$$

$$= \sum_{l \geq k} 2^l E\left[|\xi_{i_l} - \xi_{i_{l+1}}|\right]$$

$$\leq \frac{1}{2^{k-1}}$$

which implies that the subsequence $\{\xi_{i_k}\}$ converges to ξ, that is,

$$\lim_{k \to \infty} d(\xi_{i_k}, \xi) = 0.$$

Finally, according to the triangle inequality, we have

$$d(\xi_i, \xi) \leq 2d(\xi, \xi_{i_k}) + 2d(\xi_{i_k}, \xi_i).$$

Then, letting $i, i_k \to \infty$, we get

$$\lim_{i \to \infty} d(\xi_i, \xi) = 0.$$

The proof is complete.   □

7.2 Regret Minimization Model

This section introduces the regret minimization model based on the worst regret criterion (Inuiguchi and Ramík 2000; Inuiguchi and Tanino 2000). Suppose that x is a solution of the fuzzy programming model. If it is known that the fuzzy vector $\boldsymbol{\xi}$

takes value $\mathbf{c}$, then the decision-maker will have a regret as the distance between the maximum objective value and $f(\mathbf{x}, \mathbf{c})$, will be

$$\max_{\mathbf{y}} f(\mathbf{y}, \mathbf{c}) - f(\mathbf{x}, \mathbf{c}). \tag{7.6}$$

At the decision-making stage, the value for fuzzy vector $\boldsymbol{\xi}$ are clearly unknown for the decision-maker, except their possibility distributions as suggested by experts. Then the regret should be a fuzzy variable defined as

$$R(\mathbf{x}, \boldsymbol{\xi}) = \max_{\mathbf{y}} f(\mathbf{y}, \boldsymbol{\xi}) - f(\mathbf{x}, \boldsymbol{\xi}),$$

which will be called the *regret variable*, and its expected value will be called the *regret degree*.

Generally speaking, a decision-maker would like to make decision on the basis of the consideration that the regret variable should be minimized. However, since it is meaningless to minimize a fuzzy quantity, we can minimize the regret degree, and get the following expected regret minimization model,

$$\begin{cases} \min & E\left[\max_{\mathbf{y}} f(\mathbf{y}, \boldsymbol{\xi}) - f(\mathbf{x}, \boldsymbol{\xi})\right] \\ \text{s.t.} & E\left[g_i(\mathbf{x}, \boldsymbol{\xi})\right] \le 0, \quad i = 1, 2, \ldots, n. \end{cases} \tag{7.7}$$

Remark 7.1 The expected constraints can also be changed to chance constraints according to the preference of the decision-maker. In this case, a predetermined confidence level should be given.

Remark 7.2 The concepts of feasible solution, local optimal solution, and global optimal solution are given by Definitions 2.6, 2.7, and 2.8.

Remark 7.3 For simplicity, we denote

$$F(\boldsymbol{\xi}) = \max_{\mathbf{y}} f(\mathbf{y}, \boldsymbol{\xi}).$$

For each feasible solution $\mathbf{x}$, since fuzzy variables $F(\boldsymbol{\xi})$ and $f(\mathbf{x}, \boldsymbol{\xi})$ are dependent, we generally have

$$E\left[F(\boldsymbol{\xi}) - f(\mathbf{x}, \boldsymbol{\xi})\right] \ne E\left[F(\boldsymbol{\xi})\right] - E\left[f(\mathbf{x}, \boldsymbol{\xi})\right].$$

Therefore, the regret minimization model and the expected value model are not equivalent.

7.3 Fuzzy Simulation

Since the regret degree is defined by the expected value, the simulation can be viewed as a specific application of the expected value simulation algorithm. The main steps are listed as follows.

Algorithm 7.1 (Fuzzy simulation for regret degree)

Step 1. Set $r = 0$.
Step 2. Randomly generate vectors $\boldsymbol{y}_i$ and calculate the credibilities v_i for $i = 1, 2, \ldots, N$.
Step 3. Calculate $R(\boldsymbol{x}, \boldsymbol{y}_i)$ for all $i = 1, 2, \ldots, N$.
Step 4. Set $a = \min\{R(\boldsymbol{x}, \boldsymbol{y}_i) \mid 1 \leq i \leq N\}$ and $b = \max\{R(\boldsymbol{x}, \boldsymbol{y}_i) \mid 1 \leq i \leq N\}$.
Step 5. Randomly generate s from $[a, b]$.
Step 6. Set $r \to r + \mathrm{Cr}\{R(\boldsymbol{x}, \boldsymbol{\xi}) \geq s\}$.
Step 7. Repeat the fifth to sixth steps for N times.
Step 8. Return $R = a + r(b - a)/N$.

7.4 Applications

This section applies the regret minimization modeling approach to study the fuzzy portfolio selection problem. See Examples 2.4 and 2.5.

Assume that there are m stocks with fuzzy returns $\boldsymbol{\xi} = (\xi_1, \xi_2, \ldots, \xi_m)$. Then for each portfolio $\boldsymbol{x} = (x_1, x_2, \ldots, x_m)$, the regret variable is

$$R(\boldsymbol{x}, \boldsymbol{\xi}) = \max_{1 \leq i \leq m} \xi_i - \sum_{i=1}^{m} \xi_i x_i.$$

If the investor prefers to minimize the regret degree, we get the following expected regret minimization model (Li et al. 2012)

$$\begin{cases} \min & E\left[\max_{1 \leq i \leq m} \xi_i - \sum_{i=1}^{m} \xi_i x_i \right] \\ \text{s.t.} & x_1 + x_2 + \cdots + x_m = 1 \\ & x_i \geq 0, \quad i = 1, 2, \ldots, m. \end{cases}$$

Remark 7.4 Suppose that $\xi_1, \xi_2, \ldots, \xi_m$ are independent and identically distributed fuzzy variables. For each portfolio $(x_1, x_2, \ldots, x_m)$, it follows from the Zadeh extension theorem that fuzzy variables

$$\tau_i = \xi_i - \sum_{j \neq i} \xi_j x_j / (1 - x_i), \quad i = 1, 2, \ldots, m$$

are identically distributed. Since the regret variable is nonnegative, we have

$$E[R(\boldsymbol{x}, \boldsymbol{\xi})] = \int_{0}^{+\infty} \mathrm{Cr}\left\{ \max_{1 \leq i \leq m} \xi_i - \sum_{i=1}^{m} \xi_i x_i \geq r \right\} \mathrm{d}r$$

$$= \int_{0}^{+\infty} \mathrm{Cr}\left\{ \bigcup_{i=1}^{m} \left\{ \xi_i - \sum_{i=1}^{m} \xi_i x_i \geq r \right\} \right\} \mathrm{d}r$$

$$= \int_0^{+\infty} \mathrm{Cr} \left\{ \bigcup_i \{ \tau_i \geq r/(1-x_i) \} \right\} dr$$

$$= \int_0^{+\infty} \max_{1 \leq i \leq m} (1-x_i) \mathrm{Cr} \{ \tau_i \geq r \} dr$$

$$= \max_{1 \leq i \leq m} (1-x_i) \int_0^{+\infty} \mathrm{Cr} \{ \tau_1 \geq r \} dr.$$

Then it is easy to prove that the regret minimization model may be simplified to the following nonlinear programming model,

$$\begin{cases} \min & \max_{1 \leq i \leq m} (1-x_i) \\ \text{s.t.} & x_1 + x_2 + \cdots + x_m = 1 \\ & x_i \geq 0, \quad i = 1, 2, \ldots, m, \end{cases}$$

whose optimal solution is $(1/m, 1/m, \ldots, 1/m)$. That is, the optimal portfolio should invest uniformly among different stocks.

Example 7.7 Suppose that there are four stocks with fuzzy returns (see Table 3.4). We have the following regret minimization model,

$$\begin{cases} \min & E \left[\max \{ \xi_1, \xi_2, \xi_3, \xi_4 \} - (\xi_1 x_1 + \xi_2 x_2 + \xi_3 x_3 + \xi_4 x_4) \right] \\ \text{s.t.} & x_1 + x_2 + x_3 + x_4 = 1 \\ & x_1, x_2, x_3, x_4 \geq 0. \end{cases}$$

Take $N = 3000$, $G = 30$, $P_c = 0.4$, $P_m = 0.2$ and *pop-size* $= 100$. A run of the genetic algorithm shows that the optimal portfolio is

$$x_1 = 0.1440, \qquad x_2 = 0.1889, \qquad x_3 = 0.6591, \qquad x_4 = 0.0080,$$

and the minimum regret degree is 0.7769.

References

Boxer L (1997) On Hausdorff-like metrics for fuzzy sets. Pattern Recognit Lett 18:115–118

Chaudhuri BB, Rosenfeld A (1996) On a metric distance between fuzzy sets. Pattern Recognit Lett 17:1157–1160

Chaudhuri BB, Rosenfeld A (1999) A modified Hausdorff distance between fuzzy sets. Inf Sci 118:159–171

Diamond P, Kloeden K (1990) Metric spaces of fuzzy sets. Fuzzy Sets Syst 35:241–249

Fan J (1998) Note on Hausdorff-like metrics for fuzzy sets. Pattern Recognit Lett 19:793–796

Inuiguchi M, Ramík J (2000) Possibilistic linear programming: a brief review of fuzzy mathematical programming and a comparison with stochastic programming in portfolio selection problem. Fuzzy Sets Syst 111(1):3–28

Inuiguchi M, Tanino T (2000) Portfolio selection under independent possibilistic information. Fuzzy Sets Syst 115:pp 83–92

Klement EP, Puri ML, Ralescu D (1986) Limit theorems for fuzzy random variables. Proc R Soc Lond Ser A 407:171–182

Li X, Liu B (2008b) On distance between fuzzy variables. J Intell Fuzzy Syst 19(3):197–204

Li X, Shou BY, Qin ZF (2012) An expected regret minimization portfolio selection model. Eur J Oper Res 218:484–492

Liu B (2004) Uncertainty theory: an introduction to its axiomatic foundations. Springer, Berlin

Puri ML, Ralescu D (1983) Differentials of fuzzy functions. J Math Anal Appl 91:552–558

Rosenfeld A (1985) Distances between fuzzy sets. Pattern Recognit Lett 3:229–233

Index

X. Li, *Credibilistic Programming*, Uncertainty and Operations Research,
DOI 10.1007/978-3-642-36376-4, © Springer-Verlag Berlin Heidelberg 2013